Coconut Diseases

B. Srinivasulu
Principal Scientist (Plant Pathology)
Agricultural Research Station
Acharya N.G. Ranga Agricultural University
Ambajipeta - 533 214, Dist. East Godavari (AP), India

D.V. Raghava Rao
Principal Scientist (Horticulture)
Agricultural Research Station
Acharya N.G. Ranga Agricultural University
Ambajipeta - 533 214, Dist. East Godavari (AP), India

International Book Distributing Co.
(Publishing Division)

Published by

INTERNATIONAL BOOK DISTRIBUTING CO.
(Publishing Division)
Khushnuma Complex Basement
7, Meerabai Marg (Behind Jawahar Bhawan)
Lucknow 226 001 U.P. (INDIA)
Tel. : 91-522-2209542, 2209543, 2209544, 2209545
Fax : 0522-4045308
E-Mail : ibdco@airtelbroadband.in

First Edition 2007

ISBN 81-8189-188-0

Composed & Designed at :

Panacea Computers
2nd Floor, Agarwal Sabha Bhawan, Subhash Mohal
Sadar Cantt., Lucknow-226 002
Phone : 0522-2483312, 9335927082
E-mail : prasgupt@rediffmail.com

Printed at:

Salasar Imaging Systems
C-7/5, Lawrence Road Industrial Area
Delhi - 110 035
Tel. : 011-27185653, 9810064311

Acharya N.G. Ranga Agricultural University
Rajendranagar,
Hyderabad - 500 030
Phones : 24015035
Res. : 27425305
Grams : AGRIVARSITY
Fax : 91-040-24015031
E-mail: root@apau.ren.nic.in
v.c._angrau_yahoo.com
raghuvardhanreddy_s@rediff.com

Dr. S. Raghu Vardhan Reddy
Vice-Chancellor

Foreword

Coconut, *Cocos nucifera* L., is deeply immersed in the culture, religion, environment, social status and diet of millions of people of tropics. More than 93 countries grow coconut with Asia and Pacific predominating the major share of 86.3% in area and 82.5% in production. India is the third largest coconut growing country in the world with an area of 1.92 million ha producing 11,985 million nuts/annum. Andhra Pradesh is the fourth largest state in area and production in India with the highest productivity. This perennial palm, inspite of its hardy nature often succumbs to different diseases caused by fungi, bacteria, phytoplasmas and viruses.

The book, 'Coconut Diseases' written by Dr. B. Srinivasulu, Principal Scientist (Plant Pathology) and Dr. D.V. Raghava Rao, Principal Scientist (Horticulture) is an excellent contribution to the existing knowledge in coconut pathology. The book deals with the diseases of coconut caused by a number of pathogens and their management practices. The information related to all the diseases of coconut crop has been brought together in a single publication and I hope that this book would serve as a good reference and source book to plant pathologists, teachers, extension workers and students and all those involved in coconut cultivation.

I congratulate Dr. B. Srinivasulu, Principal Scientist (Plant Pathology) and Dr. D.V. Raghava Rao, Principal Scientist (Horticulture), Agricultural Research Station, Ambajipeta for their sincere efforts in bringing out a useful publication.

(S. Raghu Vardhan Reddy)

Acharya N.G. Ranga Agricultural University
Rajendranagar,
Hyderabad - 500 030
Phones : 24015011 (7 lines)
24015161 (3 lines)
Ext : 327
24015078 (Direct)
24511475 (R)
Grams : AGRIVARSITY
Fax : 091-040-24017453
E-mail: dr_angrau2yahoo.co.in
raghavpr@yahoo.com

Dr. P. Raghava Reddy
Director of Research

Preface

Coconut, an important plantation crop of Andhra Pradesh has a prime role in our day to day life. In Andhra Pradesh, it is grown in 20 out of 23 districts in an area of 1.00 lakh hectares producing 1010 million nuts annually. Among the four major coconut producing states; Andhra Pradesh commands the highest productivity of 11,003 nuts/ha in the country. However, coconut production in Andhra Pradesh is hampered by many constraints of which the biotic stresses cause significant losses. Coconut diseases, comprising a major biotic stress, pay a major role causing considerable reduction in nut and copra yields.

This book deals with the diseases of coconut caused by fungi, bacteria, phytoplasma and viruses and fulfilled a long felt need for a book exclusively on coconut diseases. All epidemiology and latest technologies in diseases management especially biocontrol of coconut diseases have been discussed in detail. It is hoped that this book would be very useful as a reference and source book to scientists, teachers, students, extension workers and farmers, who deal with coconut crop.

I convey my appreciation to authors, Dr. B. Srinivasulu, Principal Scientist (Plant Pathology) and Dr. D.V. Raghava Rao, Principal Scientist (Horticulture), Agricultural Research Station, Ambajipeta for their hard work in bringing out this book.

(P. Raghava Reddy)

Author's Preface

Coconut palm (*Cocos nucifera* L.), a tree of versatility, popularly known as 'King of Palms' and 'Kalpavriksha' has been used as a source of food, drink, oil and milk by humans and their ancestors for at least half a million years. This palm not only provides sustainable income to millions who are directly and indirectly depending on this palm but also helps to provide food security, nutritional security, employment opportunities, alleviate poverty, export earnings and import substitution.

This perennial palm, inspite of its hardy nature and adaptability to different soil conditions and vagaries of climate, often succumbs to different diseases caused by fungi, bacteria, phytoplasma and viruses.

There has been long felt need for a book exclusively on the diseases of coconut by compiling information available in the different coconut growing countries. The bibliography cited in this publication would bear evidence to the fact that the last five decades had seen the accumulation of a considerable amount of literature on different diseases of the coconut and their management including very minor pathogens of coconut.

It is hoped that this book would be useful for coconut growers, researchers, teachers, extension workers and students and also this book would serve as a good reference book on coconut diseases. The authors have tried to present up-to-date information on diseases of coconut.

The authors welcome the comments and remarks of the readers for improvement of the book in further editions.

Ambajipeta
August, 2006

B. Srinivasulu
D.V. Raghava Rao

Contents

Coconut Diseases

The coconut palm, *Cocos nucifera* Linn., with its tall, slender and uniformly thick stem and massive crown with large number of leaves, bearing bunches of nuts in their axils is one of the most beautiful and useful trees in the world (Menon and Pandalai, 1958). Inspite of its hardy nature, the coconut palm is affected by a number of diseases (Nambiar, 1994). This perennial palm though known for its adaptability to different soil conditions, moisture stress and vagaries of climate, is often subjected to attack by plant pathogens causing various diseases (Henry Louis, 2002). Anon (1979) 173 fungi associated with coconut and Brown (1973) recorded 35 fungi from coconut leaves, in India. The roots, stem and the crown regions are attacked at different growth stages of the crop, which may result in considerable economic loss to the grower. Besides fungal pathogens, bacteria, phytoplasma, virus and nematodes have also been reported to cause diseases in coconut (Menon and Pandalai, 1958; Nambiar, 1994; Srinivasulu *et.al.*, 2001a). In India, basal stem rot (*Ganoderma* wilt), bud rot, stem bleeding, root (wilt) and leaf rot are important diseases of coconut and are serious constraints to the production and productivity of the crop.

Fungal Diseases

Basal stem rot or *Ganoderma* wilt disease: *Ganoderma* spp

Basal stem rot disease is also called as *Ganoderma* wilt or Thanjavur wilt or bole rot or Anabe Roga in different states of India, caused by *Ganoderma* spp is a major constraint in coconut cultivation (Bhaskaran, *et.al.*, 1984a; Satyanarayana *et.al.*, 1985; Wilson *et.al.*, 1987; Srinivasulu *et.al.*, 2001a). In India, *Ganoderma lucidum* was first recorded in coconut palm in Karnataka state in 1913 by Butler (Petch, 1916). In Andhra Pradesh, *Ganoderma* wilt (basal stem rot) disease is caused by *Ganoderma applanatum* (Pers) Pat, and *Ganoderma lucidum* (Leys) Karst, is the major constraint in lighter soils than in black soils (Papa Rao & Govinda Rao, 1966; Satyanarayana *et.al.*, 1985; Srinivasulu *et.al.*, 2001a). The disease is also reported from Tamil Nadu (Tanjavur wilt), Karnataka (Anabe roga), Kerala, Maharastra, Gujarat and Orissa (Basal stem rot) (Bhaskaran *et.al.*, 1994; Wilson *et.al.*, 1987; Govindu *et.al.*, 1983; Briton Jones, 1940). A disease with similar symptoms has been observed in Maharashtra and Gujarat during the past few years (Bhaskaran *et.al.*, 1989). Basal stem rot of coconut caused by *Ganoderma boninense* in Sri Lanka reported by Peries (1974). *G.lucidum* (Leys) Karst was recognized as a pathogen of coconut palm (*Cocos nucifera Linn*) in the year 1846 (Petch 1916). Venkatarayan (1935) reported that root wilt of coconut, caused by *G. lucidum*, kills 15-20 palms and in severe outbreaks up to 60 trees per acre every year.

Symptoms

Peries *et.al.*, (1975); Menon and Pandalai (1958) presented a detailed description of the symptomatology of the disease. The symptoms vary somewhat and depend on the rate of progress of the disease.

Roots: The disease first starts in the root system. Initially, a few roots get infected and rot. Decay and death of the fine roots is the first underground symptom of the disease. Discoloration and extensive rotting of root system are observed which are characteristic of the disease. The rotting proceeds towards the bole. Cortical tissues disintegrate and the stele turns brown. There is a progressive reduction in regeneration of new roots. In severely diseased palms, more than 70 percent root rotting was observed. The root decay was more (up to 70%) at 0-30 cm depth than at deeper layers (up to 30%). Thus, with increase in depth from soil surface, extent of root rotting decreases (Plate-1).

Stem: Exudation of reddish brown viscous fluid from the basal portions of the stem is the first visible symptom of the disease in the affected palm. By that time, the rotting would have progressed from the bole to the basal portion of stem. The bleeding patches begin from the base and extend up to 3 meters upwards as the disease progresses. The internal tissues of the affected stem turn brown in colour and rotting in the stem can be seen up to the height of the bleeding. In advanced stages, basal portion of the stem decays completely. In some palms, the bark from the base of the stem peels off. Occasionally, some infected palms do not show bleeding symptoms. Sporophores of the fungus appear at the base of the affected trunk in palms prior to wilting or just after the death of the palm (Plate-2).

Leaves: The leaflets exhibit wilting symptoms and outer one or two whorls of leaves turn yellow. Later, they exhibit light to moderate browning followed by drooping and drying. As the disease advances, the remaining leaves also droop down in quick succession and the spindle alone remains. Under prolonged infection, the outer leaves fall off one by one, leaving only the spindle with a few unhealthy leaves around. The spindle leaves which emerge subsequently are reduced in size and do not unfold properly. Production of new leaves is delayed. Leaves break off near the base along the midrib. In certain cases, soft rot sets in the bud resulting in loss of humidity and death of cells due to breakdown of conducting elements. The affected bud emits a foul smell and in advanced stages, the crown is blown off leaving the decapitated stem. Stem shrivels and dries up (Plate-3).

Flowers: Normal development of flowers and bunches is arrested with the progress of the disease. The leaves droop down resulting in hanging down of the subtended bunches. This leads to button shedding. The nuts become barren. Most of the palms bear profusely, just prior to end at the time of initiation of symptoms. In severely diseased palms, whole nut weight, kernel weight, water content, copra weight and oil content decreases very much.

The time taken from the initial appearance of bleeding patches on the stem to the death of the palm varies from 6 to 54 months, the average being 24 months. The scolytid beetle, *Xyleborus perforans* and the weevil, *Diocalandra stigmaticollis* are found infesting the stem in large number of the bleeding patches. These insects accelerate the death of the palm. (Vijayan *et.al.*, 1973; Bhaskaran *et.al.*, 1982 & 1989; Rethinam, 1984; Bhaskaran, 1986; Srinivasulu *et.al.*, 2002). *Ganoderma* infection starts from roots of coconut and colonization was more in root portion than stem bits (Vijayan and Natarajan, 1972; Anon, 1981; Bhaskaran *et.al.*, 1982, 1984a).

Fungus nomenclature

The fungus *Ganoderma lucidum* was first described under the name Formes lucidus (Leys.) Fr. It was Butler (1909) who described the fungus in detail. The following are considered as synonyms: Formes lucidus (eys) Fr., F.amboinensis (Lam.) Polyporus lucidus Fr., P. amboinensis Fr., Polystictus egregious Masse, G. amboinensis (am) Pat., G. resinaceum (Butler and Bisby, 1913), G. sessile Murrill, Polyporus fulvellus Bres., P.resinosus Schraeder, P. curtisil (Berk) Muril and G. mangiferae (Lev.) (Van Overem, 1926).

Etiology

Aerial mycelium hyaline, thin-walled, branched with frequent clamp connection, 1.4-2 m in diameter; chlamydospores formed abundantly which are slightly thick-walled, terminal or intercalary, ellipsoid, sometimes in chains. 8.8-11.8m x 3.7-5.9m in size; cuticular cells from crustose layer hyaline to light brown, round to irregular in shape and closely packed; staghorn hyphae with projection present in some isolates while absent in others; submerged mycelium thinwalled, hyphae and chlamydospores as in aerial mycelium (Manjusri Sen, 1973; Govindu *et.al.*, 1983). The fruiting body is perennial (Plate-4), stipitate, usually lateral, sometimes sessile, corky, becoming woody later, usually 10-12 x 10-12 x 3-4 cm, but may grow up to 30 cm or more; upper surface is shining, laccate crust, ox-blood in colour, smooth. The palisade hyphae is about 40m long and is impregnated with a dark orange varnishing substance which they secrete. Hymenial surface is whitish or creamish, turning brown later, pores small, round, 90-250 in diameter. Pore tubes are about 6-7 mm long, basideiospores are brown, thick-walled, minutely verrucose, truncate at one end and 8.3-10.0 x 5.8-6.7m in size (Plate-5) (Bose, 1930; Govindu *et.al.*, 1983).

Host range

The fungus has a wide host range infecting variety of palms and several forest avenue and fruit trees. Naidu *et.al.*, (1966) reported hosts belonging to 19 families, 36 genera and 48 species. Besides coconut, the fungus infects Acacia catechu wild., A. auriculaeformis, A. cunn., A. melanoxylon, R.Br., A. nilotica (L) Willd. ex Del., Acrocarpus fraxinifolius Wt., Albizzia chinensis (Osbeck) Merr., A. lebbeck Benth., A. procera Benth., Aquillaria agallocha Roxb., Areca catechu L., Bosweilia serrata Roxb., Cassia fistula, L., C. javanica L., C. nodosa Ham., C., siamea Lam., Casuarina equisetifolia Forst., Dalbergia latifolia Roxb., D. sissoo Roxb., Delonix regia (Boj. ex Hook) Raf., Elaeis guineensis Jacq., Eucalyptus citriodora Hook., Fiscus spp., Hevea spp., Jacaranda acutifolia Humb d roxburghii Sarg., Pleiogynium cerasiferum (FVM). Parker, Pongamia pinnata (L.) Merr., Populus euramericana (Dode) Guinier, pterocarpus marsupium Roxb., Quercus semecarpifolia Smith, Shorea robusta Gaerth., Sterculia villosa Roxb., Tamarindus indica L., Terminalia alata Hayne. ex Roths. Var. nepalensiss (Haines) Fern. W.A., T. tomentosa W.A. Toona cibiata Roem, and Vitis vinifera L. (Venkatarayan, 1936; Bakshi *et.al.,* 1972; Govindu *et.al.,* 1983; Rajan, 1987; Adaskaveg and Gilbertson, 1987 & Bhaskaran *et.al.,* 1994).

Epidemiology

In Andhra Pradesh, the disease is mainly prevalent in lighter soils in the coastal districts than in heavy soils (Papa Rao and Govinda Rao, 1966; Satyanarayana *et.al.,* 1985; Srinivasulu *et.al.,* 2002a). Number of rainy days and rainfall have negative relationship with spread of basal stem rot disease in coconut (Srinivasulu *et.al.,* 2003a). Basal stem rot disease incidence was directly related to the mean maximum soil temperature, but was not related to the minimum temperature, rainfall and relative humidity (Ramasami *et.al.,* 1977, Lewin *et.al.,* 1983, Bhaskaran *et.al.,* 1985). Soil moisture stress experienced during summer months was found to favour the spread of the disease (Ramasami *et.al.,* 1977).

According to Peries *et.al.,* (1975) the disease progress rapidly in dry areas and more slowly in wet areas. The infected palms observed by them died in six to thirty months in a region with 1000 mm annual rainfall wereas they lasted for five years to six years in a region with 2000 mm annual rainfall. Soil moisture at 5, 15 and 30 cm depths had a negative correlation with disease severity. Observation made at Ambajipeta (Andhra Pradesh), showed that the disease spread was more when the range of difference in relative humidity was higher and rainfall lesser (Ramapandu *et.al.,* 1981). Presence of hard pan in the sub-soil impedes root penetration, which in turn pre-disposes the coconut palms to infection (Anon, 1976; Ramasami *et.al.,* 1977). The fungus is a soil dweller inhabiting dead as well as living plant material in the soil. It is a root parasite. The fungus attacks palms of 20-40 years of age. But incidence on palms less than 10 years old and above 40 years is also sometimes noticed (Naidu *et.al.,* 1966).

Early diagnosis of the disease

Transpiration rate (Vijayaraghavan *et.al.,* 1987), Colorimetric method (Natarajan *et.al.,* 1986), Iodine-potassium iodide staining technique (Anon., 1990), EDTA Method (Natarajan *et.al.,* 1986, Vijayaraghavan *et.al.,* 1987), simple serological techniques and ELISA (Srinivasulu *et.al.,* 2003b) etc. are suggested for early diagnosis of the disease.

Management

Several experiments were carried out over the years to control and / or manage the disease. These experiments on cultural (Bhaskaran *et.al.,* 1978; Vijayan and Natarajan, 1975; Ramadoss *et.al.,* 1985; Gunasekaran *et.al.,* 1986), Chemical (Bhaskaran *et.al.,* 1984a; Ramadoss & Bhaskaran, 1987; Ramadoss *et.al.,* 1986a; Gunasekaran *et.al.,* 1986) and biological methods (Gunasekaran *et.al.,* 1986; Srinivasulu *et.al.,* 2002b, c & 2004a, b, c) of disease control have given certain definite indications on the practices to be followed for managing the disease.

The internal symptomatology of this disease and its incidence in relation to soil microbial populations and cultural practices were reported by Bhaskaran *et.al.* (1994) that the affected plant tissues serve as source of inoculum appears to have been well recognized as most of workers suggested the destruction of infected plant materials such as old tree stumps (Steyaert, 1948; Blacklock 1954 and Rishbeth 1955). Bhaskaran *et.al.* (1994) reported that burning of brick kilns in sites of disease incidence or application of sulphur dust all over the diseased area appears to have the same effect as destruction of infected material.

To control the disease a number of attempts, such as digging of isolation trenches separating healthy and diseased palms (Butler, 1926; Venkatarayan, 1936), application of iron sulphate, disinfection of soil with carbondisulphide emulsion and formaldehyde (Narasimham, 1921) were made. Irrigation alone cannot effectively control the disease. Irrigation combined with Bordeaux mixture drenching helped in checking the disease intensity considerably (Bhaskaran *et.al.*, 1978a). Organic manure with irrigation also ameliorated the disease symptoms to certain extent. Irrigation along with farm yard manure plus burying coconut husk in circular trench around the palm plus Bordeaux mixture drenching was very effective in reducing the disease intensity (Bhaskaran *et.al.*, 1978a; Suriachandraselvan & Bhaskaran, 1999). It was also observed that burying 500 coconut husks in circular trench around the diseased palms controlled the disease (Vijayan and Natarajan, 1975; Bhaskaran *et.al.*, 1984a).

Effect of Fertilizers: Observations made at Veppankulam centre from 1977 to 1982 showed that plots treated with 350, 250 and 450 g N, P_2O_5 and K_2O respectively / palm / year had low disease index and high nut yield while higher doses of fertilizers increased the disease intensity (Bhaskaran, *et.al.*, 1978b; Bhaskaran & Ramanathan, 1983).

Effect of Micronutrients: Manganese sulphate, when applied at the rate of 227 g per palm/year, reduced the intensity of the disease (disease index 2.8 in treated palms as against 33.0 in control). The disease intensity was maximum in palms that received molybdenum (Jaganathan and Ramasami, 1975; Anon., 1978a; Bhaskaran *et.al.*, 1984a; Bhaskaran *et.al.*, 1985). However, Sindha Mathar *et.al.*, (1983) could not observe any direct effect on the disease by application of micronutrientsin Kanyakumari District of Tamil Nadu. Zinc sulphate (2%) was found inhibitory to *Ganoderma* spp (Srinivasulu *et.al.*, 2002a), while application of Sulphur and lime to the soil around the trees was found to be beneficial (Venkatarayan, 1936).

Effect of Organic Manures: Application of 50 kg farm yard manure or green leaves or 300 kg tank silt or 5 kg neem cake every year was reported to be beneficial in arresting the progress of the disease (Vijayan and Natarajan, 1975). Application of neem cake alone and in combination with drenching bordeaux mixture (1%) thrice at quarterly intervals was the most effective in reducing the intensity of the disease giving 12.30 and 11.42 disease index, respectively compared to 117.72 in control palms (Bhaskaran and Ramanathan, 1983). Field trials carried out for five years on the effect of 100 kg tank silt + 50 kg green leaves + 1 percent Bordeaux mixture soil drenchinghelped in bringing down the disease index in these plots (15.0) compared to control (75.2) (Anon., 1981b). Bakshi (1951) reported that manuring not only increases the vigour of the palm for better yields but also control the parasitium to *G. lucidum.*

Effect of Fungicides and Chemicals: Application of Bordeaux mixture during October-January was effective in reducing the intensity of the disease (Anon., 1978b, 1978c). The field trial conducted with systemic fungicides and antibiotics form 1972 to 1976 indicated that Aureofungin-sol (0.2%) was very effective in reducing the intensity of the disease (Anon., 1978a, b). Drenching with 10 lit of Benomyl (0.1%) per palm after exposing the roots also gave good disease control (Kolandaisamy and Arjunan, 1977). Drenching the soil with 40 lit of Bordeaux mixture (1%) and stem injection of Aureofungin-sol 2g plus 1 g copper sulphate in 100ml of water thrice at quarterly intervals significantly reduced the disease intensity and increased the yield of nuts (Bhaskaran and Ramanathan, 1982; Bhaskaran *et.al.*, 1984b). However, the treatment should be repeated once in three years (Ramadoss and Bhaskaran, 1987).

Anbalagan and Shanmugam (1984) reported that tridemorph at 500 ppm inhibited the growth of *G. lucidum in vitro.* Sindha Mathar and Balasubramaniam (1987) reported that soil drenching with 0.1% IBP, carboxin, tridemorph or 0.05% carbendazim in combination with neem cake at 5 kg/palm reduced the disease intensity significantly. Fieldtrial conducted at Palghat (Kerala) by CPCRI, Kasaragod showed that in tridemorph (Calixin) and Aureofungin-sol treated palms, the disease was less (Anon, 1989a). Similar results were obtained at Veppankulam Centre also, where treatments like tridemorph root feeding (thrice at quarterly interval) in combination with neem cake application and aureofungin-sol root feeding plus neem cake with and without Bordeaux mixture drenching helped in reducing disease index and increasing yield of treated palms. A similar observation was made in Andhra Pradesh also (Anon, 1990).

It was found that Aureofungin-sol (2 g) + 1% Bordeaux mixture (40 lt) + neem cake (5 kg) checked further spread of the disease in the palms (Anon, 1985). The treatment mildothane (10 g in 5 lt of water), Aureofungin-sol 1.5 g + copper sulphate 1 g in 5 lt of water (AF), carboxin stem injection 500 mg in 5 ml of water (CAR) and AF + CAR reduced the linear spread of the disease significantly when compared to control (Satyanarayana *et.al.*, 1985). Anbalagan and Shanmugam (1984) reported that tridemorph at 500 ppm inhibited the growth of *Ganoderma lucidum in vitro*. Application of Bordeaux mixture during October – January was effective in reducing the intensity of the disease (Anon, 1978). Neem cake extract, banana rhizome extract and Tephrosia purpurea root extract gave 100, 86 and 54 per cent inhibition of *Ganoderma applanatum* respectively (Bhaskaran *et.al.*, 1988).

Since the association of the scolytid beetle *Xyleborus perforans* was also noticed in some wilt affected palms, a field trial using dieldrex, heptachlor, chlordane and sulphur dust was conducted from 1966 to 1969. The palms treated with heptachlor showed minimum disease intensity (Anon, 1981b). Stem injection of 2.5 ml monocrotophos in 100 ml of water showed variable results regarding disease intensity, though the treatment marginally increased nut yield (Anon, 1983).

Manganese sulphate when applied at the rate of 227 g/palm/year reduced the intensity of *Ganoderma* (Bhaskaran *et.al.*, 1978b). Papa Rao and Govinda Rao (1966) reported that in plots treated with sulphur dust, the rate of spread was slower initially and faster after six months than in check plots. The lowest disease index was achieved with tridemorph root feeding (2%) + soil drenching (0.3%), followed by hexaconazole root feeding (1%) + soil drenching (0.2%), soil drenching with tridemorph (0.3%) and hexaconazole (0.2%) compared to root feeding alone (Naik,

2001). Fungicides *viz.*, Bordeaux mixture, Triademifon, Tridemorph, Bitertenol, Copper Oxy Chloride, Hexaconazole were found to inhibit the *Ganoderma* applanatum and *G. lucidum* under *in vitro* conditions and these fungicides were also found to be inhibitory to *Trichoderma viride* (Srinivasulu, *et.al.*, 2002b)

Effect of Botanicals: A number of plant extracts were tested for their effect on the growth of *G.lucidum.* Neem cake extract completely exihibited its growth. Banana rhizome extract and *Tephrosia purpurea* root extract gave 86 and 54% inhibition respectively (Bhaskaran *et.al.*, 1987). As banana is one of the profitable intercrops in coconut gardens and *Tephrosia* is a green manure crop raised in coconut garden, growing banana and applying *Tephrosia* in coconut basins, where the disease is endemic, may offer scope for reducing the intensity of the disease (Bhaskaran *et.al.*, 1988). Garlic extract (10%) completely inhibited the mycelial growth of *G. applanatum*, and *G.lucidum* under *in vitro* conditions (Srinivasulu *et.al.*, 2002b)

Management by Tapping 'Neera': With a view to finding out the effect of tapping 'neera' on disease intensity, tapping experiments were conducted from May to October or from September to February in palms with different disease intensities and in apparently healthy palms. In both the experiments, tapping in mildly and moderately diseased palms reduced the disease index and the effect persisted even one year after completion of tapping (Vijayaraghavan *et.al.*, 1986, 1987; Anon, 1990; Bhaskaran, *et.al.*, 1990).

Biological Control: Garrett (1955) reported that *T. viride* and *Streptomyces* spp were antagonistic to *G.lucidum.* Papa Rao *et.al.* (1976) reported that *G.lucidum* fail to grow under *in vitro* conditions, on unsterilized coconut root bits and also on sterilized roots which were contaminated with species of *Trichoderma, Aspergillus* and *Rhizopus.* Application of neem cake (5 or 10 kg per palm/year) encouraged the saprophytic soil microflora especially *Trichoderma* in coconut basins and was effective in the control of *Ganoderma* wilt (Gunasekaran *et.al.*, 1986; Bhaskaran *et.al.*, 1988; Bhaskaran, 1990). In dual culture technique, among the three species of *Trichoderma* tested, maximum suppression was noted with *Trichoderma* harzianum to the extent of 72 per cent in *Ganoderma applanatum* and 75 per cent in *G. lucidum* over control (Srinivasulu *et.al.*, 2002b). *Trichoderma viride* and *T. harzianum* were also reported to be antagonistic to *Ganoderma lucidum* (Gunasekaran *et.al.*, 1986; Bhaskaran *et.al.*, 1988; Bhaskaran, 1990, Anon, 1990). Native isolates of biocontrol agents *viz., Trichoderma viride, T. harzianum, T. hamatum, T. longibrachiatum, T. virens* and *T. polysporum* were inhibitory to the causal agents of the basal stem rot disease *viz., Ganoderma applanatum* and *G.lucidum* under *in vitro* conditions (Srinivasulu *et.al.*, 2004b). Srinivasulu *et.al.* (2001b) reported that application of 50 g T. viride + 5 kg Neem cake / palm / year treatment controlled the linear spread of Ganoderma to maximum extent (22 cm) within a period of 4 months from date of treatment application, untreated palm recorded linear spread of 776 cm. Talc formulation of 50 g *Trichoderma viride / T. harzianum / T. hamatum* in combination with neem cake 5 kg per palm per year was found effective in checking the basal stem rot disease in coconut (Srinivasulu *et.al.*, 2004b). The bacterial bioagent, *Pseudomonas fluoresense* inhibited the growth of *Ganoderma applanatum* and *G.lucidum* to an extent of 64 to 69% under *in vitro* conditions (Srinivasulu *et.al.*, 2004b). *Trichoderma* spp, *T. viride, T. harzianum* and *T. hamatum* were found to be very effective in producing volatile and non-volatile metabolites against *Ganoderma applanatum* and *G.lucidum* (Srinivasulu *et.al.*, 2004a). Field studies indicated that the talc formulations (50 g/palm) of *Trichoderma hamatum* in combination with 5 kg neem cake potentially arrested the spread of basal stem rot disease followed by *T. harzianum* and *T. viride* when fortified with neem cake @ 5 kg/palm/year (Srinivasulu *et.al.*, 2004a; 2005a, b).

The investigation concludes that management of basal stem rot disease in coconut may be possible by adopting an ecofriendly integrated disease management package (Srinivasulu, *et.al.*, 2004a; 2005a, b). The IDM components consist of

1. The diseased part of the garden should be isolated from healthy area by digging isolation trench (1 meter deep and 0.5 meter width).
2. Removal and burning of diseased and dead palms along with roots is recommended. The pit for replanting should be filled up with a mixture of soil and farmyard manure in equal quantities along with 50 g of *Trichoderma viride* talc powder + 1 kg neem cake.
3. Injury or damage to roots and pruning or cutting of roots should be avoided to prevent thechances of the soil -borne pathogen gaining entry through injured roots.
4. Since the disease is more severe in light soils with poor water holding capacity, raising & ploughing insitu of green manuring crops like Sunnhemp, Sesbania and Diancha is advised to increase soil organic matter and antagonistic microflora.
5. Selection of seedlings for new garden from *Ganoderma* infected soil should be avoided.
6. Frequent watering/irrigation should be done during summer months and while irrigation, care should be taken to avoid flow of water from diseased trees to others. Basin system of irrigation to individual palms is thoroughly recommended.
7. Application of talc formulation of *T. viride* (50 g) in combination with 5 kg neem cake/palm/year is recommended in a garden where diseased palms are noticed.
8. To manage this disease, care should be taken that even if one diseased palm is noticed in a garden, the treatment i.e. talc formulation of *T. viride* (50 g) in combination with neem cake (5 kg/palm/year) should be imposed on all the palms in the garden.
9. The biocontrol agent (*Trichoderma viride*) should not be applied in combination with fungicides especially with Bordeaux mixture.

Bud rot - *Phytophthora palmivora* (E.J.Butler) E.J.Butler

Introduction

Bud rot is a fatal disease of coconut palm, characterized by the rotting of the terminal bud and surrounding tissues. Even though it affects the palms of all ages, young palms in low lying and moist situations are more susceptible to the disease. It is generally a sporadic disease, but sometimes appears in epidemic forms. The disease commonly occurs in West and East coasts of India (Menon and Pandalai, 1958; Ohler, 1984; Srinivasan, 2001b).

Occurrence and Distribution

Bud rot is a lethal disease of coconut occurring in almost all the coconut growing countries in the world. It was first reported from Grand Cayman, an island of British West Indies, south of Cuba in 1834 (Tucker, 1926). Subsequently, it was recorded from India, Brazil, East Africa, British Guinea, Ceylon, Fiji Island, Jamaica, Mauritius, Mexico, Nigeria, the Philippines, Puerto Rico, Surinam, Trinidad, Cuba, Dominican Republic and Santo Domingo, Columbia, French Polynesia, Guatemala, Honduras, Indonesia, Ivory coast, Nicaragua, Panama, Solomon Island, United States (California and Florida), Venezuela and Vanuatu (Tucker, 1926; Butler, 1906; Menon and Pandalai, 1958; Child, 1974; Chase and Broschat 1991; Nambiar, 1994). In India, bud rot was first reported in Madras state in 1906 by Butler who named the causal organism *Phytophthora palmivora* in 1910.

The disease is sporadic in nature and outbreaks of epidemics are also not uncommon (Radha and Joseph, 1974; 1983). Incidence upto 21.3% in Andhra Pradesh, 7.0% in Karnataka, 2.0% in Tamil Nadu and 6.5% in Kerala was reported (Srinivasulu, 2006a). An outbreak in epidemic proportion was reported from Kuttiadi in Kerala leading to the destruction of thousands of coconut palms (Anon., 1996). It occurs commonly in West and East coasts of India, where coconuts are grown with rubber and palmyrah (Menon and Pandalai, 1958). Bud rot incidence was also noticed in other states like Karnataka, Maharashtra and Goa in large scale, particularly in the newly planted gardens. In non-traditional coconut growing areas of Madhya Pradesh like Bastar, Raipur and Durg also bud rot is a problem of serious concern (Dantre *et.al.*, 1997). The fungus also infects palmyra, arecanut (Menon and Pandalai, 1958) palm oil and rubber (Radha and Joseph, 1974) and cocoa, black pepper and cardamom (Das and Cheeran, 1986).

The fungus is also known to cause bud rot or similar diseases on *Borassus flabellifer* L., *Cocos campestris* Mask., *C. plumose* Hock., *Dypis madagascarensis* Nicholas., *Hivistona rotundifolia* Mart., *Howea belmoreana* Beoc., *Normanbya dactylifera* L., *Phoenix dactylifera* L., *Pinanga insignis* L., *Ptychosperma macarthii* H. Wendl., *Rhopalostylis baneri* Wendl. & Drude, *Sebal causiarum* (Cook) Becc.

Symptoms

Palms of all ages are susceptible to the disease, but it is more frequent on young palms. Briton-Jones (1940) described the disease symptoms first. The first visible symptom is the withering of the spindle marked by pale colour. The spear leaf or spindle turns brown and bends over. Basal tissues of the leaf rot quickly and can be easily separated from the crown. Spindle withers and droops down and one by one, the inner leaves also fall away, leaving only fully matured leaves in the crown. A foul smell is emitted by the rotting tissue (Nambiar, 1994). The palms succumb to the disease with the death of the spindle (Briton-Jones, 1940; Menon and Pandalai, 1958; Lingaraj, 1972). Later, infection spreads to the older leaves, causing sunken leaf spots covering the entire leaf blade spreading both up and down. Spot margins are irregular and water soaked, and when the leaves are unfolded, the characteristic irregular spots are conspicuous on the blade. In severely affected trees, the entire crown may rot (Plate-6) and in few months the trees wilt. Bud rot and premature nut fall diseases are produced on coconut. The heart leaf becomes chlorotic, wilts and collapses. The disease may spread to older, adjacent leaves and spathes, producing a dead centre with a fringe of living leaves. Light brown to yellow, oily, sunken lesions may be found on leaf bases, stipules or pinnae. Internally, the tissues beneath the bud rot discolored pink to purple with a dark brown border. Infected nuts show brown to black necrotic areas with a yellow border developing on the surface. Internally, they have a mottled appearance. Young nuts are highly susceptible and fail to mature and then fall off from the tree. Older, infected nuts ripen normally (Anon, 2000). The disease is rampant during the monsoon when the atmospheric temperature is low and the humidity is high (Muliyar *et.al.,* 1982).

The disease is called nut fall or Mahali in Kerala (Sundararaman and Ramakrishnan, 1924). The fungus also infects nuts, causing decay of immature nuts and their fall during the rainy season (Nambiar, 1994). Water soaked greyish green area develops at the stalk end of the nuts against dark green healthy area around. The lesions later turn brown and become sunken due to decay of underlying tissues. The rot extends into the husk and sometimes deep into the endosperm cavity (Plate-7) (Nambiar, 1994).

Etiology

The disease is caused by a fungus, *Phytophthora palmivora* (E.J. Butler) is a Butler. In 1906, Butler reported that the bud rot of palms in India is caused by *Pythium palmivoram,* which was later renamed as *Phytophthora palmivora* Butl (Menon and Pandalai, 1958). Veena *et.al.,* (1997) reported *P.katsurae* from Kuttiadi area in Kerala. *P. katsurae* was first isolated from Chestnut in Japan. Studies have shown the existence of two dominant species of *Phytophthora, P.palmivora* in Indonesia and Phillippines and *P. katsurae* in the Ivory Coast. Both species are found in Jamaica. *P. arecae* and *P. nicotianae* are also found in Indonesia (Renard and Darwis, 1993). Uchida *et.al.,* (1992) reported a new *Phytophthora* fruit and heart rot of coconut in Hawaii, the pathogen of which resembled *P. katsurae.* Presence of both 'Plus' and 'Minus' strains of the fungus are also seen and coconut isolates are considered as A_2 mating type (Kumar and Hegde, 1987).

The fungus mycelium is typically non-septate and hyaline, inter-cellular in the tissue drawing its nutrients through haustoria. It develops rapidly to cover the host tissue with a cottony growth, especially during highly humid conditions. The sporangiophores are simple or branched and the sporangia are pear-shaped, with prominent papillae. They are formed singly at the tips of conidiophores and are hyaline and thin-walled, measuring 38-72 x 33-42m. They germinate by releasing motile zoospores through the papillar opening. The zoospores are reniform, with two flagella and they swim for a while in water and come to rest, losing the flagella. The resting zoospores are spherical and thin walled, measuring 8-10m in diameter. Later in the advanced stages of fungal growth, oogonia and antheridia are produced, which copulate to produce oospores. The thick-walled oospores are formed inside the oogonium surrounded by its persistent wall. They are spherical and measure 35-45 m in diameter, and germinate by short germ tubes bearing sporangia at the tips. The fungus also produces thick-walled intercalary or terminal chalamydospores.

No significant differences in mycelial growth were recorded in three different media over a range of pH and temperature. A moderate sporangia formation was noted over the range of pH tested (4.0-8.0). Light had no influence on palpillate sporangia formation (Srinivasulu, *et.al.,* 2006f).

Epidemiology

Both the relative humidity and rainfall are related to infection and disease development (Radha and Joseph 1974). It was found that minimum temperature is also an important factor in disease incidence, in addition to humidity. Duration of 'favourable days' having high humidity and low temperature for extended periods determined disease development. The period elapsing between the first infection and the withering of the heart may vary from three to nine months or longer, depending on relative humidity.

The fungus might remain dormant in the leaf base during the dry season. Insect attack is considered a predisposing factor for the fungal infection. Rainfall might aggravate infection when the spotting appears on the pinnae, since the fungus will be washed down into the bud. Mortality is heavierin cold weather (Butler, 1910; Menon and Pandalai, 1958). *Phytophthora* propagules were isolated from crown debris of dead trees even after one year. This shows the potential hazard exerted by diseased trees as a source of inoculum in subsequent season. A disease forecasting equation has been developed for predicting bud rot disease in Kerala. This would help to forewarn the farmers for taking plant protection measures to check bud rot incidence (Rohini Iyer and Rasmi, 2005). In Philippines, Rillo and Paloma (1989) observed higher incidence of *Phytophthora* which was always preceded by high rainfall during the previous months. The fungus was found to survive in the frond base or basal part of the crown (Menon and Pandalai, 1958; Radha and Joseph, 1974) or in roots (Harris *et.al.,* 1984). Radha and Joseph (1976) revealed that the favourable period of infection is when relative humidity is above 94% and temperature is below 24°c. The disease is primarily spread by wind and wind blown rain (Renard and Darwis, 1993).

Bud rot disease incidence on coconut palms was low (1.6%) in sandy soils than black soils (25%) (Srinivasulu *et.al.,* 2006f).

Disease Cycle: The fungus may perpetuate in the form of oospores or chlamydospores, formed late in the season and carried over the summer months on the host debris, in crevices and natural openings of the dead tissue. With the onset of monsoon rains and favourable atmospheric temperature (18-20°C), the fungus becomes active, producing the cottony mycelium that infects the tender host tissue. With continuing favourable weather the fungus produces sporangia and numerous zoospores, which spread rapidly in the rain water falling on the tree to cause new infections. Thus, the fungus multiplies rapidly in its asexual stage, later forming the sexual oospores and asexual chlamydospores, which enable it to survive adverse climatic conditions.

Differentiation of the fungus into + and - strains has been demonstrated, showing that union of antheridia and oogonia can result in the formation of oospores, a process that can give rise to new strains. However, more studies on the strain variations of this species are needed.

Varietal resistance

Quillec and Renard (1984) observed that hybrids were less sensitive than West African Tall. Rillo and Paloma (1989) found that *Phytophthora* incidence was higher in Cameroon Red Dwarf and MYD as compared to Catigan. Coffee *et.al.,* (1990) reported that in Indonesian dwarf selections such as Jombang and Raji appeared to be more resistant to nut fall in inoculation trials. For avoiding economic loss, it is recommended that the choice of planting material should be done by taking into account the environmental conditions (Mangindaan *et.al.,* 1992).

Management

The methods of management were described by Menon and Pandalai (1958), Peiris (1962), Radha and Joseph (1974), Nambiar (1994). When the disease is diagnosed in the early stages, curative measures can save the palm. The infected tissues are removed and the wound is treated with 10 percent Bordeaux paste. The treated bud is covered (Menon & Pandalai, 1958) with polythene sheet to prevent entry of rain water, till normal shoot emerges. In case of advanced stages of infection, it is advisable to cut and remove the palms. The diseased tissues should be burnt after their removal. As a prophylactic measure, all the healthy palms in the surroundings should be sprayed with Bordeaux mixture (1%) or copper oxychloride (3 g/lit). Spraying of Bordeaux mixture (1%) to the crown as a prophylactic measure to check the disease is recommended (Menon & Pandalai, 1958; Radha, 1973). Bud rot can be controlled by application of Alliette (Fosetyl-Al) as drenching to young palms or by injections into trunks of mature trees (Mahmud *et.al.,* 1990).

Garofalo and McMillan, 1999 reported that symptomatic palms probably cannot be saved, but adjacent asymptomatic palms can be treated with Alliette (Fosetyl-Alluminium) (Fosetyl), Barrot (Ethazole) thiophanate methyl or metalaxyl. At initial stage of disease, when the spindle is just withering, application of Bordeaux paste on the crown after removing the infected tissue and a thorough cleaning prevents the spread. Treated wound should be given a protective covering till the next normal shoot emerges. Badly affected trees which are beyond recovery should be cut and burnt. All the healthy palms around the diseased palms should be sprayed with 1% Bordeaux mixture or copper oxychloride (3 g/lit) (Anon, 2000). Large scale field trials for the control of *Phytophthora* bud rot were conducted by Butler (1910) and Mc Rac (1923) in India.

The following practices are found to check the disease (Menon & Pandalai, 1958; Anon, 2000).

- Cutting and burning of badly affected trees on the spot.
- If the disease is detected in the early stages, when the heart leaf is just withering, application of Bordeaux paste on the crown after thorough cleaning and removal of infected material and if necessary, a few healthy leaves also is recommended. The wounded and the treated portion should be given a protective covering till the next emerging shoot comes out normal.
- A prophylactic spray of Bordeaux mixture to the apparently healthy palms adjacent to the diseased ones. The disease was found to be checked with spraying of Bordeaux mixture (1%) or copper oxychloride (0.3%) with 2 sprayes at 15 days internal (Anon., 2000)

Other fungicides advocated earlier against bud rot disease include Phenyl mercuric urea applied in the form of powder or pellets in the leaf axils (Peiris, 1962) or stem injection or root infusion of Aliette or Ridomil or Phosphorus acid (Brahmana *et.al.*, 1992). In copper sensitive palms, keeping perforated sachets containing Dithane M-45 in the leaf axils during rainy season is useful (Nambiar and Rawther, 1993).

Antagonistic fungi like *Trichoderma* spp, *Myrothecium roridum* and *M.verrucaria* (Tuset *et.al.*, 1992) are reported to suppress the growth of *P. palmivora* and *P. katsurae*. It has been reported that antibodies produced by these fungi may have a role in the suppression of the pathogen in the soil. Guest and Drew (2002) reported that organic soil amendments stimulate the activity of soil microbes and suppresses the survival and activity of phytophthora and phosphonate protect trees against infection. Biocontrol agents, *T. viride, T. harzianum* and *T. hamatum* and also *P. fluorescens* were found to inhibit the mycelial growth of *P.palmivora* under *in vitro* conditions (Srinivasulu *et.al.*, 2004b).

Stem bleeding: *Thielaviopsis paradoxa* (de Seynes) von Hohnel

Introduction

The disease is reported to occur in almost all coconut growing countries in the tropics. The disease was first reported from Sri Lanka in 1906 by Petch. From India it was first reported by Sundararaman in 1922. Though, the disease does not cause crop loss in the early stage, but there will be a steady decline in yield as the disease advances. In some cases, death of the palm has also been noticed. The occurrence of the disease has been reported in all soil types. However, the disease incidence is more in laterite soils and sandy soils.

Occurrence and distribution

Stem bleeding disease of coconut is a debilitating disease prevalent in all coconut growing regions in the tropics. The disease was first reported from Sri Lanka (Petch, 1906) and later reported in India (Sundararaman, 1922), the Philippines (Lee 1922), Malaya (Sharples, 1923), Andamans (Mitra, 1929), Indonesia (Renard *et.al.*, 1984; Kartika *et.al.*, 1984), Jamaica (Martyn, 1945) and in Trinidad and Tobago including Dominican Republic (Briton Jones, 1940; Goberdhan, 1961) and Ghana (Chona and Adansi, 1970). The disease has been found to occur in all soil types, but more in laterite soils and sandy soils on the sea shore or back water areas (Nambiar, 1994). A heavy incidence (10%) of stem bleeding incidence in Sri Lanka was reported by Wijesekara *et.al.,* (1999).

Symptoms

The earliest visible symptom is the presence of characteristic dark brown patches at the basal portion of the trunk. This lesion in course of time progresses both upwards and downwards. A dark reddish brown liquid exudes from the longitudinal growth cracks present on the stem bark, forming irregular streaks. These streaks become larger in size by coalition and no oozing is seen from old lesion. The gummy exudation dries up and form black encrustations surrounded by brownish orange margin (Plate-8). The tissues underlying the discoloured patches show rotting. The internal decay can be observed even in areas beyond the margins of external lesions (Plate-8). Later the tissues become black and fibrous. In young infected palms the decaying process is fast and as a result, cavities are formed from which a liquid oozes out when the bark is pressed. Severe infection results in reduction in yield and death of young palms. It has been observed that maximum symptoms of bleeding occur in June-July and the least in summer months. In the crown the outer whorl of leaves show premature yellowing followed by drooping and drying. In some cases nut fall also is noticed. In advanced stages of disease the trunk shows tapering with a reduced crown size. The crown symptoms are not conspicuous in well managed gardens receiving irrigation. A method of disease indexing for its severity has been worked out. Severe infection may lead to reduced yield and death of young palms (Ohler, 1984; Nambiar and Sastry, 1988; Nambiar *et.al.,* 1986). The affected palm shows gradual decline in yield (Ramanujam *et.al.,* , 2002).

Nambiar (1994) also reported symptoms on crown region. The outer whorl of leaves becomes yellow rather prematurely, droop and finally dry up. The trunk gradually tapers towards the apex and the crown size is reduced. Maximum symptoms of bleeding during rainy season and the least in summer months was noticed (Radhakrishnan, 1987) and disease progress was faster during months of high humidity (July to November months) and moderate temperature (Nambiar *et.al.,* 1989). Salgado (1942) reported that the bleeding patches spread spirally about half way up the stem and some time reach the crown and cause the death of palms. Papa Rao and Govinda Rao (1966) reported that up to 28 palms are affected by stem bleeding disease in AP. Sundaraman (1922) and Salgado (1942) reported that in severe cases the bleeding patches reach in to the crown and kill the palm. Sundararaman (1922) and Briton Jones (1940) observed that when infection occurs at one or more points, they would coalesce in the advance stage of the disease.

Etiology

The disease is caused by a fungus, *Thielavilopsis paradoxa* (de Seynes) von Hohnel. *Ceratocystis paradoxa* (Dade) Moreau, the perithecial stage of the fungus has also been isolated from the symptomatic tissues. The fungus has also been isolated from nut showing splitting of husk and gummy exudation. The etiology of the disease was further confirmed with the reproduction of symptom in the fungus inoculated healthy coconut trees (Nambiar *et.al.,* 1986; Srinivasulu *et.al.,* 2006a). *T. paradoxa* is also reported to be a pathogen of diseases in palmyrah, arecanut, sugarcane and pineapple (Menon and Pandalai, 1958).

T. paraodoxa produces pale brown to brown hyphae. Conidiophores are slender, arising laterally from the hyphae and produce cylindrical to oval endocondidia; when mature they are hyaline to pale brown, smooth walled (6-24 x 2-5.5 mm). Chlamydospores terminal in chains, obovate to oval, thick walled, brown, 10-25 x 7.5-20 mm. The perithecial stage is *Ceratostomella (= Ceratocystis) paradoxa* (Dade) Moreau; Perithecia partly immersed, light brown, 190-350 mm dia with numerous appendages; long and black neck, tapering up to 1400 mm, osteolar, hyaline; ascospores ellipsoid, often with unequally curved sides, hyaline, non-septate, smooth, 7-10 x 2.5 x 4 mm.

The optimum temperature for mycelial growth of *T. paradoxa* was found to be 30 °C (Nishitha Naik, 1990).

Epidemiology

The fungus is a weak pathogen and enters through wounds/ growth crevices present on palms (Ohler, 1984; Briton-Jones, 1940) subject to prolonged dry spell followed by sudden heavy rains or after heavy manuring. Damage to the stem caused by bush firing has also been found to predispose the palms to infection. Other predisposing factors are poor drainage, occurrence of hard lateritic pan, imbalance in mineral nutrients etc. Temperature also seems to play a role in symptom expression. Thus maximum symptoms of bleeding were observed in July under lateritically loam conditions with the least symptoms in summer months. Inoculation studies conducted at Kasaragod show that the progress of the disease was faster during July to November months when high humidity and moderate temperature prevail. Increased levels of P was found to cause increased disease incidence. However, no significant difference in major nutrient contents between healthy and diseased palms could be observed. No correlation was also observed between disease incidence and soil pH or electrical conductivity (Nambiar, 2004).

Management

Control measures adopted till recently consists mainly of phytosanitation involving removal of affected bark tissues with a chisel and application of hot coal tar or Bordeaux paste to protect the wound. Rahman *et.al., 1989* reported the application of coal tar after removing the infected tissues is found to be promising in controlling the disease. Control of the disease through phytosanitation and using systemic fungicides like tridemorph and carbendazim have been reported with limited success (Ramanujam *et.al.,* 1993). Lowest stem bleeding index (2.62) and the highest nut yield (59.4 nuts/palm annually) in the palms treated with tridemorph 4% (root feeding and wound dressing) along with coal tar sealing and soil application of

Gliocladium virens + neem cake + farmyard manure (Ramanujam, 1997). The chiseled tissues are destroyed by burning. Root feeding with Bavistin or Calixin (Nambiar and Sastry, 1988) and Anon, (1990), drenching Calixin (0.1%) at 25 litres per tree once in every two months (RadhaKrishnan, 1990) and root feeding with Calixin @ 6 ml in 25 ml of water (Anon, 2003) reduced the spread of disease. Fosetyl-Al, Captafol and benomyl were effective in suppressing mycelial growth of *T. paradoxa* in culture and preventing lesion development in the inoculated stem whereas chlorothalonil and metalaxyl were not effective (Bachiler and Abad, 1998). Potty and Radhakrishnan (1978) reported that increased levels of P caused increased disease incidence while, chlorine deficiency is reported to be associated with the disease and correction of chlorine deficiency is reported to have prevented the disease (Renard *et.al.*, 1984; Van Uexkull, 1985). Excessive salinity with high sodium is associated with stem bleeding (Nagarajan, 1985). Damage to stems can be checked to a certain extent by completely removing the affected tissues using a chisel and dressing the wound with hot coal tar or Bordeaux paste (Anon, 2000). Application of neem cake @ 5 kg/palm was found to increase the soil microflora, *Trichoderma* spp which were found to inhibit pathogen *in vitro* (Gowda and Nambiar, 1987) and in tissues under lab conditions (Usman, 1988). *Gliocladium virens* inoculation @ 10.0 g/pot + Neem cake (50 g/pot) + FYM (500 g/pot) showed maximum reduction in *T. paradoxa* population and soils amended with neem cake + FYM increased the levels of native populations of G. virens (Ramanujam *et.al.*, 2002). *In vitro* studies revealed that the mycelial growth of *T. paradoxa* was inhibited to a maximum extent by *Trichoderma viride* (69.35% inhibition), *T. harzianum* (62.90%), *T. hamatum* (66.13%) and *Pseudomonas fluorescens* (68.32%) (Srinivasulu *et.al.*, 2004b). Radhakrishnan and Potti (1980) reported that among the indigenous coconut cultivars screened, West Coast Tall was the most resistant.

In the early years coal tar application on the affected bark after chiseling the decayed tissues was recommended. However, this measure often does not afford good protection. Encouraging results of control could be obtained when bavistin or calixin were applied through root feeding or drenching. Since wounds on the trunks of palms predispose the palms to infection, any type of such wounding should be avoided. Application of neem cake at 5 kg/palm was found to increase the soil microflora like *Trichoderma* spp which are found to inhibit the pathogen. Summer irrigation may be predisposed in the garden to avoid extreme variations in soil moisture. Field studies indicated that application of *Trichoderma harzianum / T. viride / T. hamatum* has paste over bleeding patches coupled with soil application of bioagent @ 50 g in 5 kg neem cake has reduced the perimeter of the stem bleeding patches (Srinivasulu *et.al.* 2006a)

IDM package for stem bleeding disease (Srinivasulu *et.al.*, 2006a, b):

- Injury or damage to the stem, especially whiie ploughing should be avoided.
- Smearing of talc powder paste of *T. viride / T. harzianum / T. hamatum* on the bleeding patches on the stem (paste can be prepared by adding 5 ml of water to 10 g of talc powder).
- Soil application of 50 g of talc powder formulation of *Trichoderma viride / T. harzianum / T. hamatum* in combination with 5 kg neem cake / palm /year.
- Recommended dose of organic (50 kg FYM or green leaf manure or compost) and inorganic fertilizers (NPK 1 kg : 2 kg : 2 ½ kg) may be applied per palm/year.
- Frequent watering / irrigation should be done during summer months. While irrigating, care should be taken to avoid flow of water from diseased trees to others. Basin system of irrigation to individual palms should be adopted.
- Good drainage facilities may be provided in the garden.
- Raising and ploughing in situ of green manure crops like Sunhemp and Sesbania is advised to increase soil organic matter and antagonistic microflora.

Leaf blight or Grey leaf spot: *Pestalotiopsis palmarum* (Cooke) Stey

Leaf blight, grey blight or grey leaf spot is a widespread disease in the tropics. The disease is widespread in all the coconut growing regions and found to be particularly severe in Kerala. It was first reported from British Guyana (Copeland, 1931) and later on from Malaya, Mindanao, the Salier Island, the New Hebrides, the Bismark archipelago, Sri Lanka, India, Trinidad etc. In Jamaica, Ashby (1920) recorded a leaf disease with similar symptoms under the name 'Die back'. Horne (1908) states that the disease though never fatal but reduces the general vitality of the palms. It is reported from British Guyana and later from Malaysia, New Hebrides, Srilanka, India, Trinidad, Nigeria (Menon and Pandalai, 1958; Nambiar, 1994). Adult palms of 20-40 years of age were found highly susceptible to the disease (Karthikeyan & Bhaskaran, 1999).

Symptoms

The disease symptoms develop only on the outer whorl of leaves *i.e.*, the mature or older leaves. They appear first as minute yellow spots with a grey brown margin on the outer whorl of leaves. The spots are oval in shape measuring about 1-5 cm long (Menon and Pandalai, 1958). Gradually the centre of the spots becomes grayish white while the brown colour of the margin deepens (Plate-9). Many spots coalesce to form large irregular necrotic patches. On the upper leaf surface, the globose, spherical, rectangular or ovoid black pycnidia of the fungus are formed. In advanced stage of infection it gives a blighted appearance (Nambiar, 1994). Complete drying and shrivelling of the leaf blade are common when the infection is severe (Gour *et.al.*, 2002). The leaves in the advanced stages of the disease present a blighted appearance and hence the name 'Leaf blight' (Plate-10 & 11).

Etiology

The disease is caused by *Pestalotiopsis palmarum* (Cooke) Stey.

This is a weak pathogen. The fungus invades leaf tissues only through injuries or when the palms are growing under extremely unfavourable conditions. The organism penetrates the host when its tissues are tender and slowly establishes as a parasite (Menon and Pandalai, 1958). Ashby (1920) reported that bad drainage or potash deficiency were the factors that predispose the leaves to infection by the fungus. The fungus is considered as a weak pathogen. On the upper surface of the leaf, the black pycnidia of the fungus appear as black minute specks. The rate of conidial germination was higher (93%) on the upper surface as compared to 80 per cent on the lower surface (Lingaraju *et.al.*, 1987). The fungus needs light for sporulation. However, it sporulated in dark on a medium containing sodium chloride (Mani and Swami, 1983). On the upper surface of the leaves, the black pycnidia of the fungus appear as black minute specks. The conidia are equipped with hair-like appendages.

The fungus occurs on other palms also, especially on palmyrah palms. Abad and Magad (1977) opined that the disease incidence indicated poor nutritional status of affected palms. Imbalanced nutrition such as potash deficiency or too much nitrogen causes the seedlings to be more susceptible. Leaf damage by insects may provide an entry to the fungus and Alonzo and Palomar (1980) reported that the potassium deficiency was found to increase the blight. Many species like *P. brevista, P. gibberosa, P. stellata, P. phoenicis* and *P. versicolor* associated with leaf spots/ blight have been reported.

Management

Briton Jones (1940) reported that the disease can be managed by balanced fertilizer application, spraying of Bordeaux mixture (1%). Spraying of carbendazim, thiabendazole and thiram completely inhibited grey blight pathogen under *in vitro* and *in vivo* conditions (Das and Mahanta, 1985). Since potassium deficiency increased the blight (Alonzo and Palomar, 1980) regular application of potassium chloride reduced the disease incidence (Abad *et.al.*, 1978). None of the cultivars or hybrids were found resistant to the disease (Bhaskaran and Ramanathan,

1983), but 'Citigar' variety showed consistent tolerance to the disease (Abad *et.al.*, 1978). Removal of the older affected leaves and spraying the foliage with Bordeaux mixture (1%) would check the spread of the disease (Gour*et.al.*, 2002). Paparao *et.al.*, (1975) reported that spraying with Bordeaux mixture (1%) or Fytolan (0.5%) in May, June followed by one more spraying during November is recommended to keep the disease under check. Spraying of Mancozeb (0.3%) or copper oxychloride (0.3%) twice at monthly interval is recommended (Anon, 2000).

Tebuconazole (40 ml/20 lt) was found to reduce the disease incidence by 37% (Warwick & Abakerli, 2001). Karthikeyan *et.al.*, (2002) reported that root feeding with thiophanate-methyl or carbendazim at 2.0% reduced the disease index after three years of application to 18.96 to 20.9% respectively. Nitrogenous compounds increased the growth and sporulation of the pathogen *in vitro*, where as potassium had an inhibitory effect on the growth of *pestalotipsis palmarum* (Karthikeyan and Bhaskaran, 1998). Khalequzzaman *et.al.*, (1998) reported that carbendazium (0.1%) was an effective fungicide followed by mancozab (0.2%) against grey leaf spot of coconut. 1000 g mp/ plant to 1500 mp/plant were the most effective fertilizer levels for the control of grey leaf spot of coconut.

The best way to avoid the disease is to improve the growing conditions of the affected palms. Diseased palms may be treated with fortnightly sprayings of Bordeaux mixture of other copper fungicides or carbamates containing zinc or manganese (Menon and Pandalai, 1958; Das and Mahanta, 1985). Regular application of potassium chloride was reported to reduce the disease incidence (Abad *et.al.*, 1978).

Leaf rot disease - *Colletotrichum gloeosporioides* and *Exserohilum rostratum*

Leaf rot disease (LRD) is widely prevalent in the southern districts of Kerala and in Cumbum valley (Theni and Dindigul districts) of Tamil Nadu (Srinivasan and Rohini Iyer, 2006) and the disease generally appears on palms affected by root (wilt) disease. Leaf rot in combination with root (wilt) is a disease complex of more concen to coconut as leaf rot brings about rapid decline of the affected coconut palms (Solomon *et.al.*, 1999; Srinivasan, 1999; Srinivasan and Rohini Iyer, 2006).

Symptoms

For the first time Butler (1908) reported the symptoms of the leaf rot disease (LRD). Later several workers have described the symptoms of leaf rot (Patel, 1938; Menon & Nair, 1951; Srinivasan & Gunasekaran, 1992). Leaf rot initially appears as minute, water soaked lesions on the emerging spindle with different shapes, colours and occasionally on other young tender leaves. These lesions enlarge, coalesce leading to extensive rotting (Plate-11) especially under favourable environmental conditions *i.e.*, high rainfall, relative humidity and low maximum temperature. Mould growth (myclieal of fungi and spore masses) is also seen on the surface of affected leaflets (Srinivasan *et.al.*, 1995). The rotting progresses into the interior of the spindle and infected spindle gradually decays, emits a fowl smell and many inset-ants, maggots etc are attracted towards the decaying tissues. Tips of leaflets and midribs also often become blackish and shrivelled. The tips of rottle spindle leaflets sometimes stick together while the base of the leaflets are still open. The rotten tissues dry up, turn black and fall off (Srinivasan and Rohini Iyer, 2006).

With maturity of leaflets, the progress of rotting is slowed down in certain palms due to hardening of tissues and the basal portions of leaflets remain without symptoms, thus giving a fan-like apperanace of leaves in the crown. Even though leaf rot is non-systemic, successive infection of emerging spindles does occur and rotting occurs to a varying extent (successive infection leads to vertical spread of the disease, aided by close proximity of emerging spindles) ultimately resulting in appearance of the LRD symptoms in most or even in all leaves of the crown (Plate-12). Horizontal spread of LRD occurs from palm to palm with increase inincidence of RWD. The pattern of

LRD symptoms in whorls of infected palms is such that the symptoms expressed in entire crown in 24-43% of palms (in different soil types) implying the linear progression /vertical spread of the disease in individual palms. The younger leaves (more specifically the tender unopened spindle leaf) play a critical role in LRD incidence. The appearance of the disease symptoms on petiole, midrib/mid veins of leaflets, in natural conditions as well as upon artificical inoculations, has been also found (Anon, 1996; Srinivasan and Gunasekaran, 1992, 1998a, 2000a; Srinivasan *et.al.*, 1998). In certain palms, the digital ends of leaves break due to the disease lesions and become yellow - hang, dry and fall-off (Srinivasan, 2002a, d). In mid-whorl yellow leaves, which appear in certain RWD palms (Department of Agriculture and Fisheries, Travancore, 1938, 1939, 1940), lesions/spots does appear in leaflets that coalesce resulting in severe blighting of lamina. In such mid-whorl yellowed leaves also severe leaf rot may appear. Palms undergoing chronic infection due to the disease decline steadily and might even succumb. For indexing and quantifying leaf rot disease a grading system was evolved (Radha *et.al.*, 1961; Mathai, 1980; Srinivasan and Gunasekaran, 1996a).

Etiology

The fungi associated with leaf rot were *Exserohilum rostratum, colletotrichum gloeosphoroides (Glomerella cingulata), Fusarium solani* and *F. moniliforme* (*Gibberlla fujikurai*) (Sosamma, 2001). From the evidences accrued in recent years it is found beyond doubt that LRD is a disease of fungal complex and two fungi *viz., Colletotrichum gloeosporioides* and *Exserohilum rostratum* were main pathogens (Srinivasan, 1998, 2002a; Srinivasan *et.al.*, 2002).

Epidemiology

Increase in incidence of root (wilt) leads to increased occurrence of LRD, an air borne foliar disease. Cyclic incidence of RWD followed by infection of LRD therefore leads to development of the disease complex (RWD plus LRD) in more palms in the field (Srinivasan, 2001b, 2002d). The most severe incidence of

LRD is observed in the monsoon/rainy seasons (high atmospheric humidity and low temperature) even as disease incidence can be generally observed irrespective of seedlings. Spore population of fungi in the atmosphere is high during the months of monsoon (Menon and Nair, 1951; Radha *et.al.,* 1961; Dwivedi *et.al.,* 1979; Srinivasan, 2002d).

Varietal resistance

The varieties such as Andamon Ordinary and New Guinea were considered more resistant to LRD (Radha, 1961). In indigenous Chowghat Dwarf Green variety and in other varieties *viz.,* Andamon Ordinary, Cochin China, SSG, San Ramon and in coconut hybrids *viz.,* West Coast Tall x Laccadive Dwarf and West Coast Tall x Nyior Gading, lowest natural incidence of LRD was recorded.

Management

Control of LRD is a relief to the affected coconut as a RWD infected palm is vulnerable to fungal infection (Solomon *et.al.,* 1999; Srinivasan *et.al.,* 2001) and hence, management of LRD mainly with fungicides forms an integral part in the management of RWD complex (Koshy, 2000).

The palms treated with mancozeb @ 3 g/palm dissolved in 300 ml of water applied twice a year gave maximum reduction in disease followed by phorate (20 g) + hexazonazole (2 ml) and phorate + mancozeb and hexaconazole alone (Koshy *et.al.,* 2002).

Based on current knowledge on leaf rot, various parameters are integrated and a system evolved (phytosanitation, fungicides/chemicals, cultural/agronomic practices, disease resistance/tolerance, biological control etc.) for the disease endemic region for effective management of the disease (Srinivasan and Rohini Iyer, 2006).

- Removal of disease advanced/senile/uneconomic palms followed by replanting, if necessary, with healthy seedlings from elite palms/varieties or hybrids resistant / tolerant to the disease complex.

- Adopting all recommended packages (agronomic, cultural, practices etc) as applicable to RWD endemic region (as described elsewhere) - such practices would aid the palm to grow in sound vigour, countering the state of plant sickness.
- Regularly monitoring the individual palms, if spindle and other young leaves are found infected, the affected portions should be promptly pruned and disposed off by burning. As a follow up effective fungicide like contaf should be poured onto spindle regional (0.2 - 0.4% conc; up to one lt/palm; 2 - 3 times/year).
- Treating the palms at initial stage itself would be more useful and in moderate cases, treatment need be prolonged.
- Keeping all treated palms also under regular observations to counter any recurrence of the disease.
- General cleaning of the crown in pre-monsoon period would help reducing LRD inoculums and palms should be further monitored with a thrust during monsoons.

Minor Fungal Diseases

Alternaria Leaf Spot: *Alternaria alternata* (Fr) Keissler

In India, the disease was first reported in Andhra Pradesh (Rao and Subramanyam, 1975) and generally, the disease is seen in coconut seedlings of 1-3 years old. The disease appears as black, round or oval spots, which during summer coalesce thus forming large patches. The causal agent is *Alternaria altenata* (Fr) Keissler. The disease is also prevalent recently in coconut gardens in Northern Kerala (Nambiar, 1994). *A. tenuissima* was also associated in the diseased leaves of coconut (Mishra, 1987). The amylatic activity of diseased coconut leaves was noticeably higher when compared to healthy coconut leaves. The fungus infects *Capsicum annum* also.

Thielaviopsis leaf spot: *Thielaviopsis paradoxa* (de Seynes) von Hohnel

Ashby (1925) described a disease that was epidemic in Jamaica for some years in a coastal belt of about 60 km, with a moderate to high rainfall, occurring on young bearing palms, being 10-15 years. The disease causes large wedge-shaped spots with broad dark brown margins running across the folded segments of the young leaves in the shoot before they are pushed up into the light. As the leaves open out, the midribs wither or torn away giving the crown a ragged appearance. Protacio (1964) reported a similar condition of palms in the Phillippines. In both cases, the fungus *Thielaviopsis paradoxa* was invariably present in the rotted spots and the palms recovered naturally. In Jamaica, cutting back the central shoot below the spotted area and applying a dry mixture of lime and sulphate of copper to the severed surface cured many palms. Inoculation tests conducted in the Philippines showed its ability to cause stem-bleeding disease of coconut palms.

White Thread Blight: *Pellicularia filamentosa, P. koleroga* Cooke, *Corticium penicillatum* Petch

The disease has been recorded from Papua New Gunea, Sri Lanka, Fuji etc. and the disease is caused by the fungus, *Corticium penicillatum* Petch. The disease usually occurs on the older leaves and maybe distinguished from the ground by the presence of large dead patches on the upper surface of the leaves mainly circular in outline, comprising all or part of several to many leaflets, usually pale-tan at first but later becoming bleached to pale buff. On the under surface of the leaves, white fungal threads run in strands along the midrib while finer threads spread lengthwise over the under surface of the leaflets forming a white film. The advance threads spreading towards the tip of the leaflets resemble fans or white cotton. The fungus kills the leaf tissue and the tissue below the advancing edge of the fungus usually being green. As the disease mainly occurs on older leaves, it causes little damage.

Leaf break: *Lesiodiplodia palmarum* (Pat.) Griffon & Maubl

This disease is reported from Trinidad, Malaysia and Sri Lanka. The initial symptoms are withering and drooping of distal ends of the leaves, breaking away from the remainder of the leaves at a weak point, varying from 30-100 cm from the top, but remaining attached, hanging directly towards a pendulous section. A brown mark produced on the leaf stalk, is particularly noticeable on the lower surface. The drooping end of the leaf, being yellow at first, finally turns brown. The fungus spreads into the lower part of the leaf travelling down the leaf stalk and eventually resulting in the whole leaf becoming yellow. This disease is caused by *Lesiodiplodia palmarum* (Pat.) Griffon & Maubl and a weak parasite only attacks the coconut palms growing under unfavourable conditions. The fungus merely accelerates the death of the palms having weakened already by other causes such as lack of drainage.

Leaf stalk rot / Petiole rot: *Anthostomella cylindrospora* Booth and Shaw

The disease was reported from Sri Lanka, Malaysia, Jamaica, Papua New Guinea, New Caledonia and India. The earliest symptom is usually yellowing and withering of the tip of the leaf on the outside of the crown. The disease spreads to the adjoining leaves and finally almost all the outer leaves breakdown. Progress is slow. The stalks of the attacked leaves show dark brown, somewhat sunken spots, being 2 cm diameter, or larger on the upper and under surfaces. The lowest spots are usually situated about 10 cm above the stem. Long dark patches are also frequently present on the limbs, and gum may exude when the affected areas crack. Internally, the pots and patches show either a sudden rot with narrow red margins or are light to dark brown, depending on their age; frequently they entirely pierce through the limb. The bunches of nuts fall owing to the removal of their support and the bud is not affected. The disease is caused by *Anthostomella cylindrospora* Booth and Shaw and *A. fusispora* Booth and Shaw. Ashby (1920) isolated *Phytophthora* sp. (*P. parasitica)* from lesions on the petiole showing 'leaf stalk rot' symptoms. The disease does not cause serious damage.

Bitten leaf disease: *Thielaviopsis paradoxa* (de Seynes) von Hohnel; *Phytophthora* spp

This disease is reported from Jamaica (Nowell, 1923). It has got great similarity with the leaf rot disease. The disease derives its name from the appearance and shape of the affected leaves. The younger green unfolded leaves appear bitten off at the ends, show successive leaflets lower on the stalks as broken or reduced to stumps. Those nearest to the heartleaf show spots on successive green leaflets, which bend over and break and hang down at these points. Leaves just pushing up have black tips. Infection usually starts as a yellow discoloration with a brown edge along a narrow line beside the ribs of the leaflets and run across all the leaflets at that point. As the leaves get mature, the leaf stumps become dark and gummy.

Ashby (1917) recognized four types of 'Bitten leaf' diseases *viz.,* (1) Pine apple bitten leaf disease (*Thielaviopsis paradoxa*) (2) Hard or little leaf bitten disease (A species of yeast) (3) *Phytophthora* bitten leaf disease (*Phytophthora* spp) and (4) Rhinoceors beetle 'Bitten leaf disease'. Briton Jones (1940) however considered that these 'Bitten leaf' disease mentioned by Ashby are not specific diseases, but they are symptoms produced due to primary infection by *Phytophthora* spp and secondary infection by *T. paradoxa*. The yeast is considered to be a saprophyte following infection by the two fungi. Cutting down and burning the infected leaves and the application of Bordeaux mixture to the foliage have been advocated as control measures.

Curvularia Leaf Spot: *Curvularia* spp

Radha and Menon (1954) reported *Curvularia* spp from leaves of root (wilt) affected palms, while Subramanian in 1953 recorded *C. inequalis* and *C. palmarum* associated with dead leaves of coconut (Nambiar, 1994). In Malaysia, the disease is caused by *C.maculans* (Chan, 1974). Appearance of small, circular and yellow spots on the leaves of nursery seedlings is the characteristic feature of the disease. These spots gradually turn brown and the centre of the spot will have sunken impression. The disease is more problematic in seedlings grown under shade and the spread is very fast under conditions of high humidity (Tey and Chan, 1980). Lack of adequate nutrition and moisture stress are the other predisposing factors of the disease. Spraying of the affected seedlings with Captan or Difolatan controls the disease (Chan, 1974). Variety Malayan Yellow Dwarf is highly susceptible while Malayan Tall is resistant to the disease. A *curvularia* sp has been reported to occur in leaf rot disease complex also (Srinivasan and Gunasekaran, 1995). The above authors and Tey and Chan (1980) observed that seedlings under shade are less susceptible to infection than those without shade, but once present, the infection spreads faster under shade conditions, probably due to higher humidity.

Exosporium Leaf Spot: *Exosporium* spp

The disease was first reported by Reinking(1919) in Philippines and is caused by *Exosporium durum* (Johnston, 1965). In India, a leaf spot caused by *E. palmivorum* was reported by Mitra (1929). This occurs only in very neglected gardens and seldom causes damage. Presence of jet black wart like or black tubercular fructifications on the leaf surface is the characteristic feature of the disease. The disease generally appears only in highly neglected gardens and seldom causes damage. In some cases, the fructifications were surrounded by somewhat discoloured area of the leaf and the fungus was identified as *Exosporodium durum*.

Petiole Disease: *Anthostomella* spp

Stockdale (1907) first reported the petiole disease from Sri Lanka and from Malayisa in 1917. The disease is also recorded from Papua New Guinea and New Caledonia (Dwyer, 1937; Shaw and Booth, 1967). However, Sulladmath and Ponnappa (1978) recorded the disease from Karnataka State in India. Dark, chocolate brown, irregularly circular or elongated lesions appear on the adaxial surface of the petiole. The lesions merge to form larger ones and later spread over the entire petiole. Raised blisters in isolation or irregularly raised patches later appear. Rotting of the infected petiole tissues is noticed. Such affected petioles break at the distal end of the affected patch and the broken dry fronds hang down and the disease does not cause serious damage. The disease is caused by *Anthostomella cylindrospora* Booth and Shaw, and *A. fusispora* Booth and Shaw. A jet black stroma is produced below the surface of infected petiole. The asci are hyaline, stalked and elliptical, 100 x 21 mm and contain eight elliptical ascospores of 31 x 8.5 mm size. The ascopores are discharged when RH is 96-100 per cent at 24-28 °C by breaking of ascus wall in the locules and extruded through the pores (Sulladmath and Ponnappa, 1978).

Leaf Blotch: *Coticium* spp

The disease was found on nursery seedlings in Malaysia, particularly the MAWA and the tall coconut and to a less extent, on the dwarfs. The first symptom of the disease is the appearance of numerous small water soaked spots, mostly on the lower leaves. Initially the spots are circular with a diameter of less than 2 mm, having a light brown colour, later rapidly changing to dark brown. A few lesions coalesce, forming elongated necrotic spots and irregular blotches of necrotic tissues (Tey and Chan, 1980). A species of *Coticium* has consistently been isolated from the lesions. Several fungicides have proved to be effective under laboratory conditions. Pruning adversely affected leaves may arrest the spread of the disease and as a preventive measure, overcrowding of the nursery should be avoided.

Drechslera leaf spot: *Drechslera incurvata*

The disease is caused by *Drechslera incurvata*. The disease occurs widespread and is commonly found on palms growing under unfavourable conditions. The spots are small at first, oval and brown, enlarge and become pale buff in the centre with broad, dark brown margins. In case of severe attacks, the edges of leaves become extensively necrotic (Holliday, 1980). Experiments conducted by Gallasch (1974) with coconut seedlings showed that the severity of the disease was related to the rate of nitrogen applied. Nitrogen fertilizers increased the susceptibility of the palms to the disease whereas both potassium and phosphorus fertilizers decreased their susceptibility. Also the susceptibility of the seedlings decreased with age.

Leaf fall or leaf droop: *Phytophthora* spp

The disease manifests by the abnormal drooping of the lower leaves which gradually dry up and fall. In advanced cases, about one-third of the total number of leaves hang down disturbing the fruit bunches borne in their axils and result in premature nut-fall and button shedding. The base of the drooping leaves area swollen and present soft discoloured patches. Species of phytophthora have been detected in the discoloured portion (Dwyer, 1937).

Fruit Rot: *Phytophthora palmivora* (E.J.Butler) E.J.Butler

In India, the fruit rot disease is known as Mahali or Nut fall in Kerala (Sundararaman and Ramakrishnan, 1924). *Phytophthora palmivora* is the causal agent of the disease. Decaying of immature nuts and their fall during rainy season is the characteristic feature of this disease. At the stalk end of the nuts, water soaked grayish green area develops against dark green healthy area around. The lesion development was faster at the stalk end of the nuts. The lesions later turn brown and become sunken due to decay of underlying tissues. The rot extends into the husk and sometimes deep into the endosperm cavity if the shell has not hardened (Concibido Manohar, 1990; Quillec and Renard, 1984). The fungus *P. palmivora* was found to be MF_1, mating type A_1 in Indonesia (Bennett *et.al.,* 1986). Rate of lesion growth was inversely proportional to age, being 61.8 per cent and 39.3 percent in 4 and 10 month-old nuts and wounding was essential for successful inoculation. The fungal pathogen also causes bud rot.

The disease can be managed effectively by a prophylactic spray with Bordeaux mixture (1%) during pre-monsoon period. However, depending on the rainfall, a post monsoon spray may also be required. According to Bennett *et.al.,* (1985) the loss of production during rainy season can reach up to 75%. In Indonesia, premature nut fall is particularly severe on Yellow Dwarf (Renard and Darwis, 1993) and mainly concerned with immature bunches, particularly during the rainy season. The rate of lesion development is inversely proportional to age of the nut (Bennett *et.al.,* 1986). Bennett *et.al.,* (1986) sugget that a dipteran, *Thalostylinus* sp. probably feeding in sugary exudates of young nuts leak on the husk surface of healthy and diseased nuts could act as a vector of the fungus in Indonesia. Two species of *Phytophthroa,* namely *P. palmivora* and *P. katsurae* are important in relation to the nut fall disease (Ooka and Uchida, 1984; Steer and Coates-Beckform, 1990). The performance of coconut varieties depends on the regions where they are planted (Mangindaan *et.al.,* 1992). Rillo and Paloma (1989) claimed that red and yellow coloured nut populations are more susceptible to nut fall as compared to green coloured cultivars. Phytosanitary measures and regular plant protection measures, cultural and fertilizer application practices help prevent the disease incidence and spread. In Karnataka, immature nut fall was also found to be caused by another fungus *Botryosphaeria rhodina* (Berk & Curt) V. Arx. (Sulladmath and Shantappa, 1980). In Karnataka, immature nut fall was also found to be caused by another fungus *Botryosphaeria rhodia* (Sulladmath and Shantappa, 1979).

Lethal Bole Nut: *Marasmiellus cocophilus* pegler

One of the important diseases of coconut now observed sporadically, often causing seriouseconomic losses. The disease was first noticed in Kenya and Tanzania and was reported to reach epidemic proportions in several areas along the coasts of these countries (Bock *et.al.,* 1970). Later the disease had been reported from Solomon islands and Madagascar. The disease mainly affects young palms in the field or seedlings in the nursery. The characteristic diagnostic symptom of the disease is the presence of day, reddish-brown rot at the base of the bole. Around the decay, there is a distinct yellowish margin, 0.5 to 1.0 cm wide. The symptoms are moré pronounced in two to four-year-old seedlings. In the seedlings, the presence of white mycelium was noticed in cavities in the area of rot. However, in mature palms, mycelial colonization proceeds rather slowly. The crown shows symptoms of general wilt and leaves turn yellowish. In order palms, the outer whorl of leaves droop leaving an erect tuft of wilted, short leaves which unfold before emerging fully. A foul odour is emitted from the inner whorl of leaves as a result of soft rot in the tissues.

The casual fungus is *Marasmiellus cocophilus* pegler (Bock *et.al.*, 1970). Sporophores occur on exposed roots and leaf bases at ground level in seedlings. The coconut debris in soil is also colonized by the fungus. Bock *et.al.*, (1970) found that the fungus gained access only through wounds in the roots. *Marasmiellus* spp have been known to cause preemergence shoot rot of coconuts in Malaysia (Tey and Chan, 1980) and in Solomon Islands (Jackson and Firman, 1982). The important species are *M. inodrma, M. palmivora, M. cocophilus, M. albofuscus, M. crinisequi* and *M. semiustus. M. inoderma* colonies on the husk, infects the embryo and invades the cavity (Jackson and Firman, 1979). *M. palmivorus* was found to attack germinating nuts in Malaysia causing 20 to 30 per cent losses (Tey and Chan, 1980).

Dipping seednuts, after partial husking at the top, for 15 minutes in 250 ppm phenyl mercury acetate was recommended to control preemergence root rot (Jackson and Firman, 1982). Bock *et.al.*, (1970) suggested the following control measures (a) Selection of seedlings in nurseries and subsequent transplanting should be as early as possible (b) Seedlings should be transplanted very carefully and damaged root ruined and disinfected (c) Seedlings should rot be taken from affected areas. (d) Periodic soil sterilization of nurseries (e) No cultivation should be done between palms, especially where disease is present.

Anthracnose: *Colletotrichum gloeosporioides penz (Glomerella cingulata* (Stonem.) Spadulding & Schrenk)

The disease is also known as fruit rot disease of coconut and is caused by *Collectotrichum glosporioides* Penz (Almeida Jr. and Aquino, 1978). The disease was found to affect the harvested nuts to a tune of 1.5 to 41% in Brazil. However, *C. cingulata* and *C. paucisetum* are known to cause leaf spot in India (Mitra, 1929; Johnson, 1965). Symptoms include development of dark grey to brownish black lesions on the fruit with an irregular margin. These lesions later coalesce and cover large parts of the fruits. Finally the affected nuts are desicated and deformed and finally show longitudinal ruptures causing rotting of internal parts. Nut splitting due to *Colletotrichum* infection has been reported in Kenya (Menon and Pandalai, 1958). The infection usually starts in the softest part of the fruit. Fruits, being attacked, show longitudinal ruptures, allowing other organisms to enter, causing rot of the inner part of the fruit (Ohler, 1984).

Spear rot: *Fusarium* spp

'Shoot rot' disease, as popularly known, is mostly seen in nursery seedlings in India (Sundararaman, 1925). The disease is also reported from Malaysia (Tey and Chan, 1980). Dark brown stripes are noticed on the spear or the next leaf. Later, the lesions traverse downwards very rapidly. This is followed by the spear decay and the spear snaps at this point on a slight pull. However, the bud remains intact and *Fusarium* spp is the causal agent of this disease. The fungus, *Diplodia* spp has also been reported to cause rotting of the coconut seedlings (Leefmans, 1934). Management practices to control this disease include phytosanitation, avoidance of over crowded nurseries and application of copper fungicides or Carbendazim as effective agent against the disease. A *Fusarium* species is always associated with the disease. Pathogencity tests showed this to be the casual agent. In Indonesia, Diplodia sp. has been reported to cause rotting of seedlings (Leefmans, 1934). MAWA and yellow dwarfs are fairly susceptible to the disease. The infection is rarely fatal but it severely retards seedling growth (Tey and Chan, 1980). Control may be obtained through sanitary practices and by avoiding overcrowding of the nursery. Application of copper fungicides or carbendazim were found to be effective against the disease.

Immature nut fall: *Phytophthora* sp

Dropping of the buttons of the pistillate flowers before and after fertilization is a common feature in coconuts. However, an abnormal shedding of a large number of buttons may be due to fungal infections or physiological or environmental factors (Gadd, 1923). Nutfall or 'Mahali' has been reported by Sundararaman and Ramakrishnan (1924) from Malabar. Immature nuts are shed with their stalk ends discoloured. The discolouration may extend up to the point of attachment of the nut to the inflorescence axis. Kernel of the affected fruit is rotten. Presence of *Phytophthora* sp. on the discoloured patches and their inoculation trials proved that the fungus is identical with *P. palmivora* var *arecae* (*P. palmivora*) which causes 'Mahali' or 'Koleroga' disease of arecanuts (Sundararaman and Ramakrishnan, 1924).

Nutfall is an associated symptom of several diseases like leaf rot, root rot and bud rot. Johnston (1911) reported that in the incipient stages of bud rot, the falling of immature nuts, staining of the flower spikes and the dying and falling over of the middle undeveloped leaves is a common feature (Menon and Pandalai,1958).

The view that the shedding of buttons is due to attack of pests was first advanced from BritishSolomon Islands (Dwyer, 1937). The moth (Acritocera negligens), the beetle (Diocalandra taitensis), the coreid bug (Amblypelta cocphaga) and rats are associated with this trouble (Menon and Pandalai, 1958). Premature fall of nuts is considered to be caused by physiological factors depending on soil and weather conditions *i.e.*, inadequate soil moisture, excessive moisture, water logging or the deficiency of any of the major nutrients (Menon and

Pandalai, 1958). Stockdale (1921) has described serious nut fall in Sri Lanka when there is an excessive rainfall and in heavy soils. Gadd (1923) reported that water deficit following prolonged drought caused considerable fall of the female flowers or buttons and shedding of buttons also caused by deficiency of nutrients.

Root infections: *Fomes* spp

Bryce (1924b) in New Guinea observed palms affected by *Fomes lignosus* and *F. lamaoensis (F. noxius)* Corner. Petch (1910) and Small (1927) recorded a 'Root disease' of coconut from Sri Lanka. Petch isolated *F.lucidus* from the butt portion of affected palms and Small found *F.lignosus, F.lucidum (Ganoderma lucidum)* and *F. applanatum* in combination with other root infecting fungi associated with the disease. The earlier symptoms are observed in the older leaves which start drooping and withering and remain suspended around the trunk for several months before they are shed. Younger leaves remain green for some time. The tree becomes barren due to the suppression of the inflorescence. The crown is reduced in size and the new leaves produced are successively smaller and yellowish in colour which finally wither as the bud decays. *F. lignosus* infections occur mainly on palms planted on lands newly cleared and where old stumps have been allowed to remain. The disease popularly called 'brown root disease' is associated with the infection of *F. noxius* in the bole of the palm. The bole when cut open shows a honey combed appearance, pale yellowish brown in colour and is light and tindery. Control measures includes adoption of sanitary measures to eliminate the infection. Root rotting is not likely to be serious where the palm is vigorous and healthy and hence soil conditions favourable for the palm maybe maintained (Menon and Pandalai, 1958).

Root infections: *Rhizoctonia* spp and *Pythium* spp

Ashby (1921) recorded in the diseased roots of palms affected by 'Die-back', which is a deficiency symptom, the occurrence of *Rhizoctonia* and *Pythium.* Reduction in the nutrient demand of the tree by the removal of immature nuts and the regeneration of roots by the application of blue stone (copper sulphate) around the base of the trees, restore the trees to proper health. In Florida, *Thielaviopsis paradoxa* was recorded by Giddings and Wood (1925) as causing root rot of seedlings. Thompson (1924) reported from Malaya 'a new disease' considered to be a root disease evidently transmitted by root contact, since its spread was checked by digging isolation trenches.

Post Harvest Diseases

Copra quality is found to be reduced by many fungi and important of them attacking copra are *Aspergillus niger, A.glaucus, A. flavus, A. fumigatus* and *A. ochraceous* (Cooke, 1932; Ward, 1937), *Ceratostomella adiposum* (Thompson, 1933), *Botryodiplodia theobromae, Aspergillus spp,* and *Rhizopus oryzae* (Rao *et.al.*, 1971) and *Penicillium frequentas* (Nair and Nathan, 1971). *Endosporostilbe* sp. (Patil and Kelkar, 1975) has been observed recently (Nambiar, unpublished) on the copra and shell of coconut. *Penicillium citrinum* (Susamma and Menon, 1983) and *Drechslera hawaiiensis* (Sharma *et.al.*, 1985) are also known to attack copra. The fungus *D. hawaiiensis* enters the fruit while attached to the bunch itself. The fungus, *Aspergillus* spp enters into the copra meat under high moisture conditions and thereby causing discolouration and also an increase in free fatty acid levels. Susamma and Menon (1983) reported that copra having 5 per cent moisture and kept in gunny bags lined with polythene sheet remained moderately free from infection. The fungi, *B.theobromae, Aspergillus* spp and *R.oryzae* enhanced the cellulolytic enzyme activity in the affected copra (Susamma, 1980). They also found that monsoon periods favoured the development of both bacterial and fungal populations. Subramanyam 1965 and Paul *et.al.*, 1980 observed frequent association of *Bacillus* spp with copra spoilage while others (Paul *et.al.*, 1980) noted that bacterial population was replaced by fungi when the temperature rose and moisture content of copra comes down. Srinivasulu *et.al.*, (2003c) observed the association of *Aspergillus flavus, Pencillium* spp, *Aspergillus niger, Rhizopus spp, Drechslera* spp , *Botryodiplodia* spp and other *Aspergillus* spp as the common inhabitants of copra during storage. Effective management of copra spoilage by mycoflora include application of biocontrol agents viz, *Trichoderma viride, T. harzianum* and *T. hamatum* which are highly inhibitory to the copra mycolflora under *in vitro* conditions (Srinivasulu *et.al.,* 2006d). Prevention of copra spoilage during storage can be effectively done by application of preservatives like Menadione (1000 ppm), Potassium meta bi sulphate (2000 ppm) and benzoic acid (2000 ppm), which are highly inhibitory to the aflatoxin producing copra inhabitant *i.e., A. flavus.*

The other minor fungal diseases reported on coconut are:

Bitten leaf - *Ceratocystis paradoxa* (Dade) C. Moreau (anomorph: *Chalara paradoxa* (De Seyn.) Sacc.)

Bipolaris - *Bipolaris incurvata* (C. Bernard) Alcorn leafspot

Black scorch - *Ceratocystis paradoxa* (Dade) C. Moreau (anomorph: *Chalara paradoxa* (De Seyn.) Sacc.)

Bud rot - *P. heveae* A. Thompson; *P. katsurae* Ko & Chang; *P. nicotianae* Breda de Haan; *Fusarium moniliforme* J. Sheld; *F. solani* (Mart.) Sacc.; *Graphium* sp.

Catacauma leaf spot - *Catacauma mucosum*

Damping-off - *Fusarium* spp; *Phytophthora* spp; *Pythium* spp; *Rhizoctonia solani* Kühn

Dry basal rot - *Ceratocystis paradoxa* (Dade) C. Moreau (anomorph: *Chalara paradoxa* (De Seyn.) Sacc.)

Ganoderma butt rot - *Ganoderma boninense* Pat.; *G. tornatum* (Pers.) Bresad.; *G. zonatum* Murrill

Graphiola leaf spot - *Graphiola phoenicis* (Moug.) Poit.

Koleroga - *Phytophthora arecae* (Coleman) Pethybridge

Leaf blight - *Cytospora palmarum* Cooke

Leaf spots - *Alternaria* sp; *Botryosphaeria disrupta* (Berk. & M.A.

Curtis) Arx & E. Muller; *Capitorostrum cocoes* K.D. Hyde & Philemon; *Cercospora* sp; *Curvularia lunata* (Wakk.); Boedijn; *Cylindrocladium pteridis* F.A. Wolf; *Drechslera gigantea* (Heald & Wolf) Ito; *Drechslera halodes* (Drechsler) Subram. & Jain; *Epicoccum nigrum* Link; *Helminthosporium* sp; *Macrophoma* sp; *Macrosporium cocos* Pass; *Melanconium* sp; *Mycosphaerella palmicola* Chaudhury & P.N. Rao, emend; *Periconiella cocoes* M.B. Ellis; *Pseudoepicoccum cocos* (F. Stevens) M.B. Ellis; *Phomopsis* sp; *Phyllosticta palmetto* Ellis & Everh; *Ramularia necator* Massee

Lixa grande - *Sphaerodothis acrocomiola* (Montagne) Von Arx & Muller (*Coccostroma palmicola* (Speg.) Von Arx & Muller)

Lixa pequena - *Phyllachora torrendiella* (Batista) Subi Leau

Nut fall - *Phytophthora arecae* (Coleman) Pethybridge; *P. palmivora* (E.J. Butler) E.J. Butler; *P. heveae* A. Thompson; *P. katsurae* Ko & Chang; *P. nicotianae* Breda de Haan; *Fusarium moniliforme* J. Sheld., *F. solani* (Mart.) Sacc.; *Graphium* sp.

Powdery mildew - *Oidium* sp.

Queima das folhas - *Botryosphaeria cocogena* Subileau; (anamorph *Lasiodiplodia theoboomae* (Pat.) Griffon & Maudl).

Root rot - *Fusarium* spp.; *Phytophthora* spp.; *Pythium* spp.; *Rhizoctonia solani* Kühn

Stem bleeding - *Chalara paradoxa* (De Seyn.) Sacc.

Stigmina leaf spot - *Stigmina palmivora* (Sacc. in Trel.) S.J. Hughes

Algal leaf spot - *Cephaleuros virescens* Kunze

Bacterial Diseases

Bacterial leaf blight

Bacterial leaf blight has been reported from New Britain and New Guinea (Dwyer, 1937). It is a serious disease on coconut during wet seasons. The significant symptom of the disease is the drying up of the leaves. The worst affected palms present the appearance of severe drought-affected trees. The infected leaflets may appear either at the base of the petiole or along the whole length of the rachis. Dark coloured lesions are found on the back of the midribs of the affected leaflets. These are usually large and irregularly margined, their edges being raised, appear yellowish in colour, turning in course of time to a lighter green water soaked zone. In advanced stages, the diseased areas become mottled in appearance and gradually become more pitted and stand out prominently. Often confusion arises between the bacterial blight and leaf spot (blight caused by *Pestalotia palmarum*). However, the lesions caused by *P. palmarum* are smaller and are more or less with a regular margin. The dark pustules of *Pestalotia* infection are absent in the bacterial disease. To manage the disease adapting the method of cutting off and burning the affected materials is considered to be effective.

Bacterial bud rot – *Erwinia* spp

Bacterial bud rot caused by *Erwinia* spp was also reported (Chase and Broschat, 1991).

Phytoplasmal Diseases

Root (wilt) disease of coconut

Root (wilt) disease (RWD), a non-lethal disease but debilitating malady of coconut palm was first reported from Kerala state (India) more than a century ago (Butler, 1908; Pillai, 1911; Varghese, 1934). Presently, reported to be prevalent in different degrees of intensity over a contiguous area of 4,10,000 ha causing an estimated annual loss of about 968 million nuts (Anon., 1985; Anon., 1986).

Symptoms: The diagnostic external symptom of the disease is the abnormal inward bending of the leaflets termed ribbing or flaccidity (Plate-13). Foliar yellowing of the outer whorl leaves and marginal necrosis are the other associated symptoms (Radha and Lal, 1972). In seedlings and juvenile palms, yellowing of foliage is virtually absent and flaccidity is the only symptom evident (Menon, 1937). The root system in the early stage of disease is apparently normal with few rotton roots. However, with the progress of the disease, extensive rotting of roots is a common symptom (Menon and Pandalai, 1958). The main roots and rootlets start drying from tip backwards (Menon and Nair, 1951). The capacity of the palms to regenerate fresh roots is drastically reduced (Michael, 1964). Shedding of immature nuts, drying up of spathes and necrosis of spikelets from tip downward in unopened inflorescenceis noticed in certain cases (Menon, 1937). A higher percentage of pollen produced is either sterile or with low viability (Varkery and Davis, 1960).

Internal symptoms include general stunting of epidermal cells, reduction in thickness of the cuticle on the adaxial surface and differential rate of division of the upper epidermis in the longitudinal and transverse plains in leaves of diseased palms resulting in slight downward curling of leaflets. Similarly, reduction in wall thickness of cells, sclerenchymatous fibres and bundle sheath was also observed in diseased palms. The distribution of stomata per unit area was found to be more in leaflets of diseased palms (Joseph and Shanta, 1963). Internal browning of root vascular elements extending up to the cortex and at times accompanied by disintegration of vascular tissues in about 60% of the roots of diseased palms was also observed (Indira and Ramadasan, 1968). However, Dwivedi *et.al.,* 1978 using an antioxidant in primary fixation of tissues failed to demonstrate vascular browning in externally healthy roots of disease affected palms. Reduction in thickness of the walls of the metaxylem elements resulting inuneven shape of the xylem component, developmentof tyloses, increased chromophily and necrotic obliteration of phloem are also observed in roots of diseased palms (Govindankutty, 1981).

Etiology: The etiology of the disease remained unsolved until recently. A number of abiotic factors and biotic agents were implicated as the cause of the disease.

Abiotic factors

Physiological factors: Earlier investigations attributed the disease to certain physiological factors, either the lack of nutrients, excess or inadequate water, disturbing the physiological process of the palms and thus predisposing to the disease. Varghese (1934), after carefully examining the various factors associated with the disease, had concluded that the disease is not physiological but fungi, virus and certain toxic substances produced by microorganisms living around the root zone of the palms might contribute to the disease.

Derangements in the water absorbing capacity of roots in diseased palms (Anon, 1970) so also the uptake and transport of water through the trunk (Ramadasan, 1964) and permeability changes in leaf and root tissues resulting in leaching of metabolites (Anon, 1977) are noticed. It has also been found that the mechanism of stomatal regulation is adversely affected in diseased palms. The stomata in diseased palms failed to close in response to soil and atmospheric drought resulting in excessive water loss (Rajagopal *et.al.,* 1986). Thus, impairment caused to both the facets of water transport namely, absorbtion and transpiration lead to internal water stress.

A higher respiration rate (Michael, 1978), derangement in translocation and distribution of sugars (Mathew, 1977), an altered nitrogen metabolism (Varkey *et.al.,* 1969) with an increase in amino acid concentration in leaves particularly of arginine (Pillai and Shanta, 1965), an accelerated phenol metabolism with a fall in phenol content and an increase in phenol oxidizing and synthesizing enzymes (Joseph and Jayasankar, 1973, 1979; Joseph *et.al.,* 1976; Joseph, 1983) are the other changes observed. Such malfunctions encountered in diseased palms are rather suggestive of a pathogen mediated altered host metabolism than of a physiological disorder. Physiological and biochemical changes observed in diseased palms such as derangement in uptake and translocation of solutes, higher respiration and transpiration rate, altered nitrogen metabolism and an accelerated phenol metabolism are indicative of a pathogen mediated aberrant host metabolism that of a physiological disorder (Mathew *et.al.,* 1991).

Soil and nutritional aspects: The impact of soil conditions and associated nutritional factors on the disease incidence has been on investigation since 1939. Based on a survey and study of the disease in relation to the soil conditions in the erstwhile state of Travancore, it was concluded that the disease affected soils are generally poor in major nutrients especially potash and the soils are mostly acidic with less content of exchangeable bases and poor base exchanging capacity (Menon and Nair, 1949; Sankarasubramoney *et.al.,* 1954, 1955, 1956; Pandalai *et.al.,* 1958a, b). An accumulation of major nutrients and silica in the leaves of diseased palms was also noticed. However, no perceptible difference in micronutrientlevel was evident and cadmium and strontium toxicity was also ruled out (Verghese *et.al.,* 1962a, b). Detailed analysis of soil and tissue samples from healthy and diseased tract of all the coconut growing area in Kerala has ruled out the possibility of the direct involvement of any major nutrients in the incidence of the disease (Pillai *et.al.*, 1975; Pillai, 1975). Cecil (1975) observed the Ca and Mg contents of healthy palms significantly higher than those of apparently healthy and diseased palms. He also observed that palms in the root (wilt) affected areas are in a state of unbalanced nutrition as compared to healthy palms. However, fertilizer field trials with the application of different levels of major and secondary nutrients, Ca and Mg could neither control the disease nor prevent fresh incidence although a general improvement in yield was observed (Menon and Nair, 1951; Chettiar *et.al.*, 1959; John *et.al.*, 1959; Nair and Radha, 1962; Lal, 1966). The role of micronutrient in the incidence of disease has also been ruled out (Davis and Pillai, 1966). Soil sickness as a contributing factor of the disease was also excluded based on field fertility trials and analysis of tissue and soil (Cecil and Amma, 1991).

Biotic agents

Fungi: Bourdillon in 1906 first reported that the disease is 'Fungoid' in nature (Varghese, 1934). Butler (1908) on isolation of *Botryodiplodia* spp from the roots of diseased palms suggested that root rot alone could be sufficient to cause the disease syndrome. Further studies resulted in the isolation of two species of *Rhizoctonia solani* Lutu and *R. bataticola* (Taub) Butl. Besides *Botryodiplodia theobromae* pat. (Menon and Nair, 1949) from the rotting root system of diseased palms. Pathogenicity experiments with *R. solani* and *R. bataticola* in the field as well as under pot culture conditions, although brought about rotting of roots, yet failed to produce the foliar symptoms characteristic of the disease (Menon and Nair, 1951). In subsequent investigations, *cylindrocarpon effusum* Bugn. and *Fusarium equiseti* (corda) sacc. were isolated from apparently healthy roots of diseased palms (Joseph, 1978). *C. effusum* and *C. lucidum* were also isolated from burrowing nematode lesions of the roots (Sosamma and Koshy, 1978). However, pathogenicity experiments with *C. effusum* and *F. equiseti* singly and in combination on coconut seedlings grown in individual field tanks failed to reproduce the symptoms of the disease (Anon, 1985b).

Bacteria: Menon and Nair (1951) reported the isolation of two bacteria from roots of diseased palms. Srivastava *et.al.,* (1969) observed characteristic streaming movement of bacteria in the vascular tissues of roots of diseased palms and isolated a bacterium tentatively identified as Pseudomonas sp. Employing an enriched medium, isolated a bacterium identified as *Enterobacter cloacae* (Jordan) Hormaeche and Edwards from the stellar portions of root tips of diseased palms. However, inoculation of coconut seedlings with *E. cloacae* failed to produce the symptoms of the disease thereby ruling out its involvement in the incidence of the disease (Jayasankar and George, 1991).

Nematodes: Possible involvement of a soil borne pathogen or a pathogen transmitted through soil-borne vectors warranted nematological studies. Investigations on the nematodes present in coconut roots and soil around resulted in the identification of 35 genera of phytonematodes including species of Xiphinema, Longidorus and Trichodorus known virus vectors and the burrowing nematode Radopholus similes (Weischer, 1967; Mathen, 1969; Mathen *et.al.,* 1970; Khan *et.al.,* 1971). Weischer (1967) opined that low population density of nematodes, their widespread occurrence and the general distribution pattern of the disease could exclude the nematodes from being considered as primary cause of the disease. Considering the notoriety of the burrowing nematode, Radopholus similes as having caused great havoc in citrus and banana in other countries, its role in the etiology of the disease was assessed in a pathogenicity experiment. Coconut seedlings in field tanks inoculated with up to one million *R. similes* failed to reproduce root (wilt) symptoms even after four years of experimentation. Extensive surveys covering root (wilt) diseased and disease-free areas also could not bring out a correlation between its presence and occurrence of root (wilt) disease (Anon, 1988).

However, pathogenicity experiments with these organisms (fungi, bacteria and nematodes) singly and in combination failed to reproduce the symptoms of the disease thereby indicating their non-involvement in the disease causation (Joseph and Lilly,

1991; Jayasankar and George, 1991; Sosamma and Koshy, 1991).

Virus or Virus like organisms: Nagaraj *et.al.* (1954) based on the systemic nature of the disease and resemblance of symptoms to other known plant virus induced diseases (yellows type) proposed a viral etiology for the disease. This theory gained further significance with positive transmission of the disease to coconut by sap inoculation and through the insect vector *Stephanitis typical* under field condition (Nagaraj & Menon, 1956) and under insect proof condition (Shanta *et.al.,* 1964).

Cotyledonary leaves of cowpea plants mechanically inoculated with sap of diseased palms produced malformation and crinkling in emerging trifoliate leaves. Hence, this was claimed as an indicator host for the presumed 'virus' (Shanta and Menon, 1960). The physical properties of the 'virus' was worked out employing cowpea as the test plant (Shanta and Menon, 1961). In view of the peculiar nature of symptoms on cowpea and lack of proof on its passage through bacterial filters, Holmes (1965) suggested that although the pathogen resembled a virus, it might as well be a related organism like a spirochaete or sporozoa. The symptoms on cowpea was found to be very inconsistent in subsequent studies. After detailed investigation, it was concluded that cowpea cannot be used as a reliable test plant (Sasikala and Pillai, 1978). Summanwar *et.al.,* (1969) reported the isolation of a virus from roots and leaves of diseased coconut palm. Based on positive serological reaction to three different strains of TMV antiserum, it was identified as a strain of TMV. Maramorosch and Kondo (1977) reported the presence of icosahedral virus-like particle of 56 nm diameter in epidermal and ground parenchyma cells of diseased palms. These particles were later identified as plasmodesmata sectioned in tangential plane (Parthasarathy, 1978). Polyacrylamide gel electrophoretic analysis of isolated nucleic acids from diseased palms did not indicate association of viroid either (Randles and Hatta, 1980). However, Shanta *et.al.,* (1975), through detailed electron microscopic, pathogenicity studies through serological studies, have ruled out the association of TMV with root (wilt) disease of coconut.

Phytoplasma: Solomon *et.al.*, (1983) observed the presence of mycoplasms like organisms (MLOs) in sieve tubes of juvenile tissues such as sub meristem, petiole of developing leaves, rachilla of unopened inflorescence and root apices of diseased palms. Tissues from healthy palms are free of such organisms. The prokaryotes are phloem bound and observed only in sieve tubes. The mollicutes are delimited by a trilamellar unit membrane and contained DNA strands and ribosomes. Pleomorphic forms varying from circular to oval and occasionally beaded or filamentous ones are observed. The size of the cocoid forms ranged from 250-400 nm (Plate-14) (Solomon, 1991). In sieve tubes the phytoplasma generally occupied a parietal position close to the sieve tube wall or seen in proximity to sieve plates (Solomon, 1997). Distribution of the phytoplasma within the vascular bundle is rather sparse and not all sieve elements in a patch contained them. Some of the bundles are totally free of the organism. In the older leaves moribund forms lacking internal contents only were observed (Solomon and Geetha, 1993). The apparent absence of phytoplasma in mature parts and their occurrence in tender tissues suggest the movement of the organism in the phloem assimilate to the 'sink region'. None of the other biological agents reported to be associated with the disease earlier could be observed in the vascular tissues (Solomon *et.al.*, 1999). The possible association of trypanosomatid flagellates was also studied by examining rechilla of inflorescence of six palms each in the early, middle and advanced stages of disease. The protozoan flagellate was not observed in any of the samples (Anon., 1991).

Dienes' staining (Deeley *et.al.*, 1979; Razin, 1983) and fluorescene staining using the fluorochrome 4', 6 diamidino-2 phenyl indole 2 Hcl (DAPI) (Seemuller, 1976) that binds with nucleic acid component of phytoplasma is used for detecting phytoplasmal infection in coconut, test plants dodder laurel and insect vector, exhibited abnormal bluish colouration indicating presence of phytoplasma. Such staining sites are absent in healthy tissues. Positive staining reactions with DAPI or Hoechst 33258 showed fluorescing area in extra nuclear sites indicative of the presence of phytoplasma and are more frequent in junctions of vascular bridges and also closed to the periphery of the sieve tubes (Solomon and Govindankutty, 1991).

Transmission: Phytoplasmas with the exception of a member of the Pentatomorpha of the heteropterous group are generally transmitted by insects belonging to the Homoptera group. Record of insects on coconut in India did not enlist any belonging to this group. Hence an inventory of insects on coconut was made using various traps and through direct examination of coconut foliage. This led to the identification of a leaf hopper *Sophonia greeni,* a plant hopper *Proutista moesta,* both belonging to the Homopteran group, besides lace bug, *Stephanitis typical* (tingidae) already known (Rajan and Mathen, 1984 & 1985). The vector role of lace burg and plant hopper have been conclusively established (Mathen *et.al.,* 1990; Anon., 1996). Detection of phytoplasma in pariwinkle, dodder laurel and source palm confirmed transmission of the disease (Sasikala *et.al.,* 1988).

Disease resistance / tolerance: Field evaluation of 45 cultivars and 62 hybrid cqmbination is in progress in the institute farm and cultivators gardens since 1972 to locate genotypes resistant/ tolerant to root (wilt) disease. None of the cultivars / hybrids showed the desired level of resistance / tolerance to the disease. However, the hybrid, Chowghat Orange Dwarf (COD) x West Coast Tall (WCT) under ideal management yielded higher number of nuts compared to WCT palms of identical age in the early years of production (Bavappa *et.al.,* 1986). The percentage of disease incidence was also comparatively less than the Talls. Chowghat Green Dwarf (CGD) showed field tolerance of over 90 per cent to root (wilt) disease (Anon., 1972).

Management: One of the significant features of the disease is that it is not lethal but a debilitating malady which responds to ideal management. Two strategies, one for the heavily diseased contiguous area, and another for the mildly affected area have been formulated. In the heavily diseased area, the yield of palms can be sustained or even improved through adoption of integrated management practices - removal of disease advanced and juvenile palms, balanced fertilizer application, addition of organic manures, raising of green manure crops in the basins and incorporation, irrigation during summer months, practicing, plant protection methods and adopting inter and mixed cropping. Mixed farming in the diseased gardens involving the raising of fodder crops in the inter spaces, maintaining milch cows and recycling of organic waste has aided in increasing the yield of palms by 28.1% (Sahasranaman *et.al.*, 1983). Mixed cropping with cocoa increased the yield by 27-35% and slowed down the decline of palms (Amma *et.al.*, 1982). Similarly cultivation of tapioca, elephant foot yam and yams in the interspaces of palms in the disease affected gardens for a period of three years increased the nut yield by 4.96 to 15.57 and 8.07% respectively (Menon and Nayar, 1978). The strategy for the mildly affected areas is to contain the disease by removing all the diseased palms. Following the eradication of a few diseased palms at Shenkotta (Tamil Nadu) and at Nadathara (Kerala) in 1971. Recurrence was not noticed subsequently (Anon., 1980). Since, there are no curative or prophylactic measures available to combat the disease the long term option is to breed for disease resistance. Concurrently adoption of recommended management practices could yield economic returns per unit area. Systematic rouging of diseased palms in the mildly affected area could prevent the further spread of the disease (Anon., 2005).

Tatipaka disease: Phytoplasma

Tatipaka disease is a unique disease on coconut reported in Andhra Pradesh and the disease is first identified in 1959 in Tatipaka village of Razole taluk in AP and named after the village Tatipaka (Rajamannar *et.al.*, 1994). The disease is generally observed on the palms in the age group of 20 to 60 years while young palms below 20 years are rarely seen affected (Rao *et.al.*, 1956; Narasimhachari, 1990).

Symptoms: Symptoms of the disease were thoroughly studied by Rao *et.al.*, (1956); Pandu & Rajamannar (1983) and Rajamannar *et.al.*, (1994). In the initial stages of disease development, the root growth is vigorous and as disease advances, the number and size of the roots are generally reduced and coupled with extensive root rot. Tapering of the stem below the crown is most common. Sometimes, diseased trees may exhibit twisted trunk or trunk bent in 'S' shape below the crown. The leaves are reduced in size and number and they become light green and slowly turn to yellowish green. Subsequently, all the leaves in the crown turn yellow with marked reduction in the size of the leaflets (Plate-15). In some cases, the leaflets of some leaves may adhere together without normal splitting, giving fasciated appearance to the leaf. On some trees, the leaflets may develop water soaked chlorotic spots. The fronds give a bow-like appearance due to bending at the middle, instead of bending at the tips as in normal trees. The dried up fronds are usually found hanging below the crown. The patches are very small with less number of rachillae. The diseased palms produce less number of bunches year after year. In early stages of the disease, bunches contain normal nuts but reduced in size. As the disease severity increases, the bunches contain either a mixture of normal and atrophied nuts or only atrophied nuts. In extreme cases of disease, the trees do not produce even the atrophied nuts and become barren. The normally hard mesocarp turns soft forming depressions when pressed with the thumb. At this stage, the nuts contain low quantity of water and copra and these nuts do not stand storage. As the disease advances, the nuts get shrivelled and atrophied. The atrophied nuts do not contain water and copra and even the shell is absent and nuts are very small with full of fiber only. At times, some nuts show longitudinal cracks with occasional gum formation. Size of the crown is reduced with pale green foliage and can be recognized even from a distance. In severe cases, the crown appears like that of a date palm. Finally the tree sheds leaves in slow succession and may die.

Etiology: Presence of Phytoplasmas in Tatipaka diseased samples was first reported by electron microscopy (Plate-16) by Randles and Hatta, (1980) and by Srinivasulu *et.al.* (2006c) by light and fluorescent microscopic detection techniques (Rajamannar *et.al.*, 1994).

Management: Several experiments were conducted at ameliorating the condition of the affected palms by manures and micronutrients (Subbaiah and Rao, 1963), chemicals (Rajamannar *et.al.,* 1994) but no recovery was noticed. Tatipaka disease been almost has eradicated from Andhra Pradesh by proper strategies (Rethinam *et.al.,* 1989) and mostly the efforts made by implementing eradication programmes by conducting palm to palm survey (Rajamannar *et.al.,* 1994) involving removal and destruction of and by avoiding affected trees, nuts from infected tree for nursery and raising seedlings only in disease free gardens. Avoid transport of seedlings from infected areas (Anon, 2000).

Lethal yellowing (LY) - Phytoplasma

Lethal yellowing is a fatal malady reported in the western part of Jamaica and Florida. In the past two decades, more than 50% and 80% of coconut palms in Florida and Jamaica have been killed by this lethal disease (Tsai, 1982). Disease symptoms are characterized by total nut fall, leaf yellowing, necrosis, blackening of opened and developing inflorescences and ultimate rotting of the spear leaves and growing points, results in the death of coconut palm phytoplasma were discovered in phloem tissues of lethal yellowing diseased palms (Plavscic Banjac *et.al.,* 1972). Howard *et.al.,* (1983, 1984) could establish the role of plant hopper myndus crudus Van Duzee as the vector of the disease. Tsai (1982) recommended an integrated management approach to manage the disease which includes eradication, quarantine chemotheraphy, development of resistant palms and vector control.

Lethal yellowing like diseases of coconut

Cape St. Paul Wilt Disease (CSPWD): The most pressing problem facing the coconut industry in Ghana is a Lethal Yellowing type disease called Cape St. Paul Wilt Disease (Dery *et.al.,* 1997). The CSPWD is caused by a phytoplasm (Tymon *et.al.,* 1995) and is suspected to be transmitted by insects. The use of tolerant ecotypes is the most promising and economic method of bringing the disease under control.

Lethal yellowing type diseases of coconut were also reported with different nomenclature namely kanicope disease. In Togo, Awka wilt in Nigeria, Kribit disease in Cameroon and Lethal disease in Tanzania, Kenya and Mozambique (Neinhaus *et.al.,* 1982; Schuiling *et.al.,* 1992; Dery *et.al.,* 1997). These were the different strains of the lethal yellowing phytoplasmas around the world (Tymon *et.al.,* 1998)

Coconut stem necrosis: This disease is prevalent in North Sumatra and Peninsular Malaysia. The green colour of the spear leaf changes to brown without a marked transmission - yellowing symptomatic of LY. Extensive necrosis occurs within the stem. The palm dies in about five months after appearance of the first symptom. Phytoplasmas are attributed as causal agents of coconut stem necrosis (Turner *et.al.,* 1978).

The other phytoplasmal diseases reported on coconut are :

Awka disease - Phytoplasma

Blast - Phytoplasma suspected

Cape St. Paul wilt - Phytoplasma

Cedros wilt - *Phytomonas stahellii*

Heart rot - *Phytomonas stahellii*

Kaincope disease - Phytoplasma

Kalimantan wilt - Phytoplasma suspected

Kribi disease - Phytoplasma

Lethal decline - Phytoplasma

Lethal disease - Phytoplasma

Pudricion del cogollo - Phytoplasma

Root wilt disease - Phytoplasma suspected

Stem necrosis - Phytoplasma suspected

Virus and Viroid Diseases

Coconut Cadang-Cadang Viroid (CCCVd)

'Cadang-cadang', locally meaning growth failure or running out disease, is a highly destrutive disease reported from the Bicol province of the Philippines (De Leon, 1951; Calica and Bigornia, 1953). The disease is rarely seen in palms before they commence flowering but, after this, incidence increases more or less linearly with age, with a rate of increase in the number of diseased palms ranging from 0.1 to 1% per year. Areas of high incidence may become low incidence over a 20-30 year cycle (Randles *et.al.*, 1988). The disease results in the premature decline and death of coconut palms. Total losses of about 30 million palms and annual yield losses of about 22,000 t of copra have been attributed to Cadang-cadang disease in the Philippines (Zelazny *et.al.*, 1982). The symptoms of the disease have been described in detail by Zelazny *et.al.*, 1982; Hanold and Randles, 1991a).

Symptoms

Early stage (lasting 2-4 years): Yellow leaf spots appearing water soaked in reflected light, translucent yellow in transmitted light. Nuts become small and rounded, with characteristic equatorial scarifications.

Medium stage (lasting around 2 years): Leaf spots become numerous, giving the lower two-thirds of the crown a yellowish appearance. Inflorescences become necrotic, infertile and nut production ceased. Frond production and size decline.

Last stage (lasting around 5 years): Leaf spots almost confluent. Whole crown yellow/bronze coloured and much reduced in size and number of fronds. Leaflets become brittle and palm disease. Time from appearance of first symptoms to tree death ranges from around 8 to 16 years and is generally greater in older palms (Zelazny *et.al.,* 1982; Hanold & Randles, 1991a).

Etiology: Infectivity is associated with naked, low molecular weight, single-stranded, predominantly circular, viroid ribonucleic acid. CCCVd contains a region of around 44 nucleotides which is common to most viroids (the central conserved sequence). The viroid is presumed to have a natural rod-like structure including a partially double stranded region comprising 13 based pairs, with single stranded loops (Haseloff *et.al.,* 1982). CCCVd has a number of molecular forms; the minimum infectious form is 246 nucleotides in size, but an additional cytosine may be inserted at position 197. A variable region exists at the right hand end of the molecule, at which reiteration of either 41, 50 or 55 nucleotides can occur, producing large forms of 287 to 301 nucleotides. These are unique features of CCCVd, as all other known viroids have only been detected as a single molecular form. Dimeric forms with the same sequence variations as their respective monomers are also associated with the disease, but are always isolated in smaller amounts than their monomers (Randles *et.al.,* 1988). A detailed description of the sequence and structures of CCCVd is given by Haseloff *et.al.,* (1982.)

Host range: Coconut, *Corypha elata* and oilpalms (*Elaeis guineensis).* In addition, arecanuts, *Chrysalidocarpus lutescens,* dates, *Ptychosperma macarthurii, Roystonea regia* and *Veitchia merrillii* are susceptible to inoculation with CCCVd (Anon, 1985; Imperial *et.al.,* 1985). CCCVd like sequences have also been identified in some symptomless herbaceous monocotyledonous species growing near CCCVd infected coconuts (Hanold & Randles, 1991a).

Management: No known control measures exist for CCCVd in the field and specific control recommendations cannot be developed until the epidemiology of CCCVd is more clearly understood. Potential control strategies include elimination of reservoir species, vector control, mild strain protection and breeding for host resistance. Eradication of diseased plants is usually performed to minimize spread but is of dubious efficacy due to the difficulties of early diagnosis.

Coconut foliar decay

This disease is caused by a virus which affects coconut ecotypes introduced into Vanuatu. The middle and lower leaves of affected palms turn yellowish and brown, dries up later resulting in the mortality of the palm. The vector of this disease is a plant hopper myndus taffani Bonfils (Julia, 1982).

Other Virus and Viroid diseases

Tinangaja - Coconut tinangaja viroid

Foliar decay - SS-DNA, identity uncertain

Miscellaneous Diseases and Disorders

Sooty mould

Sometimes, a large portion of the coconut leaves is covered with a black layer of fungus growth. The fungus develop on the sweet exudation of some insects such as aphids. Where spraying is possible, a cheap effective and harmless mixture of 225 g soft soap, 4.5lt water and 9lt kerosene may be used. The water should be heated and the soap dissolved in water after which kerosene is added and stirred in the mixture until an emulsion is obtained. For spraying, the emulsion may be diluted to 1 : 8 or 1 : 9 mixture with water and for fogging, pure emulsion is used (Ohler, 1984).

Crown choke

The disease was first reported from Assam in 1964 (Chakraborthy *et.al.*, 1970). It was then called 'Crown rot'. This disease was later reported from West Bengal, Kerala and Tamil Nadu. Disease with similar symptoms has been reported from Indonesia, Ivory Coast, New Guinea, Sri Lanka and the Phillippines. Reinking (1961) and Ohler (1984) stated that 'Bristle top' disease has similar symptoms. The first symptom of the disease is the emergence of shorter leaves with deformed and crinkled leaflets. The leaflets are associated with severe tip necrosis and their number decreased progressively when the attack is acute. The affected leaflets do not unfurl and in many cases give a choked appearance to the frond. The tip of the leaves become hooked; the hooks mostly seen in the terminal pinna; although they may occur in any portion of the frond (Chakraborthy *et.al.*, 1970; Baranwal *et.al.*, 1989). As the disease progresses, a severely necrotic black, sticklike leafstalk devoid of any leaflet emerges. The outer whorl and leaves look healthy and remain green. In the due course, the inner leaves crowd around the bud and as a result normal unfurling of the flag leaf is prevented. In advanced stage, necrosis of the primordial tissues takes place and the crown dies. The affected palms do not die suddenly, but it slowly losses vitality and finally succumbs within 3-4 years. Premature nut fall has been also noticed. The inflorescence emergence is hindered resulting in gradual decline in yield and finally the affected palms does not produce any nuts. No pathogenic fungi or bacteria were reported associated with the disease. Application of borax at the rate of 50 g per palm during post-monsoon period is beneficial in improving the condition of the palm (Chakraborthy *et.al.,* 1970) and palms in the early stage of disease responded better (Anon., 1989).

Pencil point disease

The disease was reported from Burma, New Guinea, Sri Lanka and India (Menon and Pandalai, 1958). The disease is also called 'Tapering stem wilt' (Jaganathan and Ramasami, 1977). Jaganathan and Ramasami (1977) had recorded up to 40 per cent incidence in some gardens in Tami Nadu. Important symptoms include reduced vigour, stunted growth and lanky appearance. The older leaves become chlorotic first to be followed by other leaves in succession. The leaves of affected palms are reduced in size, the number of fronds and spathes will be reduced. The trees in the advanced stage of the disease cease to produce any nuts. There is a gradual reduction in the girth of the stem towards the tip presenting the appearance of a pencil point. In course of time, the growth ceases and the palms succumb to the malady. However, no pathogenic organisms were reported to be associated with the disease. Application of micronutrients like boron, manganese sulphate and zinc sulphate along with NPK fertilizers help in alleviating the disease.

Hart Rot

The disease is prevalent in Surinan, Ecuador, Brazil and Trinidad and is caused by the trypanosomatid flagellate *Phytomonas stahelii*. The disease is transmitted by the pentatomid bug *Ochlerus sp*. The trypanosome vectors of oilpalm hart rot namely the pentatomid bugs *Lincus lethifer* and *incus tumidifons* were found to transmit the pathogen to coconut (IRHO-CIRAD Activity Report 1991).

Red ring disease

Red ring disease described as an endemic disease in South and Central America has been reported from Brazil (Joffilly, 1948), Venezuela, Columbia, Panama and British Honduras (Martyn, 1953). The first symptom is a progressive yellowing and browning of the leaves. The early stages of progress are rather indistinguishable in appearance from the natural dying off of leaves due to drought, water-logging or poor condition of growth. The difference becomes apparent by the process of discolourationwhich soon affects the healthy leaves and spreads to those in which the leaflets are not even fully expanded. At this stage, the tree is conspicuous by its rich colouring of brown, orange and yellow in all but the central tuft of leaves. These conditions may develop within the first three weeks of infection. The tree starts shedding immature nuts. Secondary infection by bacteria sets in among the softer tissue at the apex of the stem (Nowell, 1919; Fenwick, 1968; Salas, 1980; Kastelein, 1987). The causal organism is the nemotode *Rhadinaphelenchus cocophilus* (Blair and Darling, 1968). A cross section of the effected stem of the diseased palms shows a typical orange coloured ring usually above 3.5 cm wide about 5.0 cm inside the periphery. The palm weevil *Rhynchophorus palmarum*. L is the vector of red ring nematode (Hagley, 1965, Griffith, 1968).

When a tree dies, the soil in the neighbourhood becomes heavily infested with the parasite. Hence, affected trees should be cut down, the roots gathered and the whole lot burnt. Digging of isolation trenches to check the spread of the nematodes from palm to palm controls the disease to a certain extent. Since nematodes are highly resistant to chemical poisons, soil disinfection is not efficient. Based on the theory that the worms travel on the outside of the stem it is considered possible that a band of crude oil on the stem may hinder their ascent. The tolerance to infection in coconut by the nematode seems to increase with age of the palms (Blair and Darling, 1968).

False bud rot or lightning injury

A case of sporadic bud rot termed 'False bud rot' was reported by Sharples (1925) and injury due to lightning strike also was reported (Dwyer, 1937). False bud rot is characterized by the central shoot and the bud being affected only after the outer leaves have died, followed by rotting of bud, gradual loss of vitality, crown grows sparse and thin and allows an easy entry of bud rotting organisms. The central leaves area affected last, with all the other leaves collapsed and a single central leaf remaining, stiff and upright and the 'bud tissue' still healthy. The roots die off without showing noticeable symptoms. The affected palms are usually present in a circular area around the focus of the lightening strike. Serious affected palms usually die first, while the surrounding palms succumb slowly due to the delayed effect and other secondary factors. The injured palms have a dried up appearance. The dried up fronds often have the petioles split transversely which hang down loosely. Bleeding of the stem may also be noticed, with and without the association of borers and *Thielaviopsis* sp.

Wilt diseases of unknown origin

'Wilt' in the broadest sense includes all plant diseases, the symptoms of which are expressed by the premature withering of the leaves and the flower bunches and the gradual loss in vitality of the plant, which is invariably its reaction to the failure in water supply. However, certain specific palm diseases are designated as wilt diseases *viz.*, 'Tapering stem wilt', 'Bronze leaf wilt', etc.

Tapering stem wilt

Tapering stem wilt otherwise known as 'Pencil point disease' was observed in several of the coconut growing areas. It was reported from Burma (Furtado, 1923), Ceylon (Bryce, 1921), Trinidad (Briton Jones, 1928) and Jamaica (Briton Jones, 1940). The characteristic symptoms of the disease are the yellowing of the tips of the leaves, a reduction in their size, a gradual decrease in the girth of the trunk and a dwindling in size of the crown. The diminution in the size of the leaves and the diameter of the system progresses until the small crown fails entirely to produce new leaves and the palm dies. The trunks are hard and the vascular bundles are reddish in colour. From Ceylon, Cooke (1950) reports two types of 'Tapering stem wilt'. One is the gradual tapering described above and another which is more rapid and designated 'Quick tapering'. In the later case the yellowing of older leaves and the paling of the younger leaves which are the earliest symptoms are accompanied by the unsymmetrical nature of the crown and the development of black band just below the crown. Other symptoms observed are similar to those described earlier. Further examination of the tapering portion revealed that it was soft, black, corrugated and about a meter long. The growing point was small and found rich in potash and there was extensive root decay.

Bronze leaf wilt

The typical symptom of the disease, as the name signifies, is the bronzing of the leaves and their wilting and subsequent shedding. In a healthy palm, the lowest and the oldest leaf including the stalk, on aging, turns to yellow and later bronze or reddish brown colour and is defoliated. Usually in a normal palm, one or at times two of the oldest leaves show such yellowing and discolouration prior to shedding. On the other hand, in a palm affected by bronze leaf wilt, more than two leaves show yellow and bronze colouring coupled with the yellowing at the tips of the next two or three leaves. Wilting progresses in absolute regularity in the order of age from the oldest to the youngest. The discolouration sets in at the tip of the leaf and extends backwards to the base. Before the youngest leaf is affected, the cabbage is killed by bacterial rot which starts at the base of the leaves and spreads inwards to the cabbage. Bronzing of the leaves is accompanied or preceded by the shedding of the nuts; the spathe also turns brown from the tip backwards towards the base slightly in advance of the discolouration of the tips of the leaves, in the axils of which the spathes are borne (Briton Jones, 1940).

Internally a reddening of the ground tissue towards the centre was found in rare instances by Park (1932) and Briton Jones (1940) observed a general reddening of the ground tissue. The experimental evidence revealed that environmental factors leading to a rapid water deificit is considered to be responsible for inducing the disease (Martyn, 1934; Bain, 1937; Briton Jones, 1940). The adverse soil conditions include water logging, too open soils or soils which drain off quickly, or an impermeable strata below the surface soil which lead to a shallow root system and result in effects similar to drought. Briton Jones (1940) is of the view that the trouble might be due to 'Physiological drought'. Improvement in soil conditions which will maintain an equilibrium between root absorption and leaf transpiration during the dry season will tend to check the disease. Recovery due to watering although temporary, has been recorded by Briton Jones (1940) and Bain (1937). Bain (1937) also recommends watering in conjunction with application of potash and ammonium sulphate.

Maturation wilt

'Maturation wilt' which is of a temporary nature is distinct from the more permanent and serious wilt diseases of the mature palms reported from New Guinea (Dwyer, 1937). Pronounced tip withering occurred on the centre leaves of the affected palms, and some dying back was also seen on the outer leaves. On the back of the petioles (leaf stalks), pale brown streaks were seen. These symptoms showed up near the tip of the leaf or near the base, but, as a general rule, commenced about two-third of the way up from the base. The scattered lesions, although moist in appearance, were quite firm, but later they tended to spread and link up, becoming sunken below the general level of the outer surface of the mid rib. Eventually the diseased portions split longitudinally, exposing the fibers after the ground tissues had fallen away, and the unhealthy parts become quite dry and hard. A cross-section made after drying out had occurred showing a pale, sulphur yellow layer immediately underneath the sunken tissue; this merged into a brown discolouration extending through the petiole almost to the inner surface.

In longitudinal section, the discoloured portion showed up as full reddish-brown streaks. Such rotted portions were seen in some cases to extend below the fibrous tissues at the base of the leaves, but in no case did it appear to kill the palms. A tip-wither of the leaflets which extended from the top of the fronds almost to the base generally occurred concurrently with the stem-rotting mentioned. Around the edges of the withered sections on the leaflets a distinct halo was noticed in some cases and the 'die-back' from the tips way very distinctive. Pale areas with brown margins also appeared in the centre of the leaflets. The above symptoms only showed up on the fronds comprising the crown, while the older leaves were usually not affected. Generally the affected palms are tall, very erect often presenting a frond-chocked appearance. Only bearing palms and those at the flowering stage are affected. The nuts from the diseased palms show an unhealthy, yellow colour. Sporadic bud rotting develops in severely affected palms.

The fungus *Pestalotia palmarum* and the insect *Brontispa froggatti* were scattered widely in the plantation and the disease was more intense where a heavy infestation of whitefly, *Aleurodicus destructor* and unidentified scales were present. However, no virulent pathogenic organism was associated with the diseased specimen. Detailed investigations indicate that the disease is localized and physiological in origin. The adverse soil conditions prevalent in the areas where the disease occurs were considered the causal factor, although the disease was also observed in palms growing in soils well suited to coconuts. The appearance of the disease was associated with a pronounced spell of dry weather following heavy rains. Insects attack and infections by weak pathogenic fungi are considered to aggravate the disease, as with the incidence of abundant rainfall and the adoption of sanitary measures the palms recovered. Sudden wilting of coconut palms under certain changed soil conditions has been noticed (Pandalai *et.al.,* 1954).

Other wilts

Stockdale (1907) reported a disease of coconut from West Indies designated as root disease caused by *Lasiodiplodia* (*Botryodiplodia theobromae*). The symptoms observed are the wilting and yellowing of the leaves which begins at the tips and gradually spreads all over the leaflets. The leaves remain suspended for a considerable time beyond they are shed. Immature nuts are gradually shed and the flowers produced later do not set. The cortical tissues of the root are being destroyed (Petch, 1910).

A wilt disease was reported from Jamaica (Anon, 1908). Wilting and yellowing of the older leaves, petioles break, leaving the stumps of the wilted leaves on the crown. Nuts are shed gradually and the female flowers produced subsequently do not set. The stem of the diseased tree develops red discolouration. The cortical region of the roots is disorganised and the roots are found to harbour *Botryodiplodia.*

Petch (1910), Small (1927) and Park (1928) reported a disease from Sri Lanka as 'root disease' of seedlings and other palms. Withering and drooping of the outer leaves which remain hanging for a long time, reduction in yield by the shedding of the immature nuts and suppression of the flowering bunches. *F. lucidus* = *Ganoderma lucidum,* (Petch, 1910; Park, 1928) *Diplodia* sp, *Poria* sp, *Fomes* spp, *Macrophomina phaseolina* (Small, 1927) were reported to be associated with the disease. Frogatt (1914) from New South Wales reported a wilt disease affecting the bearing palms causing rapid death of the palms within three or four months. A wilt disease of coconut with the external symptoms of desiccation of the leaves from the base upwards and of the pinnae from the tip downwards was reported from Mauritius. Occasionally *Botryodiplodia* was found in the roots of the affected palms (Ṣhepherd, 1926).

Deformity growths

Head droop: Head droop is otherwise known as 'Cork screw' or 'Cabbage droop' or 'Strangle disease' is reported from New Guinea and New Ireland (Dwyer, 1937; Simmonds, 1924). The top of the palm bends over and in extreme cases the whole stem makes a semi-circular growth. The leaves appear collected and twisted forming distinct rosette witha few old leaves sharp and dialing back. Finally the whole stem taking a distinct loop or 'S' curve. During the period of droop, no spathes are developed. The cause of the disease has not yet been established.

Frond choke: *Sporadic* appearane of frond choke from New Guinea reported by Dwyer, 1937. The terminal fronds tend to stand upright and their growth is very poor. The older leaves at the periphery are crowded around the bud with their bases held together tight instead ofremaining free. The palm presents a sickly yellow appearance and appears to lack vigour. The etiology of this disease is still under investigation.

Other miscellenous diseases and disorders of unknown etiology reported are:

Dry bud rot - Etiology not known, but possibly vectored by the insects *Sogatella kolophon* and *S. yubana*

Frond rot - Physiological disorder

Bristle top

Finschafen disease

Leaf scorch decline

Malaysia wilt

Natuna wilt

Premature decline

Soccoro wilt

Bibliography

Abad, R.G. and Magad, S.S. 1977. Effect of KCl fertilization on coconut leaf spot and yield. *Philip. Phytopath.* 3 : 74-77.

Abad, R.G., Gallego, V.C. and David, J.R. 1978. Reaction of some coconut plantations to leaf spot disease of the nursery stage. Philipp. J. of coconut studies 3 (3) : 21-28.

Adaskaveg, J.E. and Gilbertson, R.L. 1987. Infection and colonization of grapevines by *Ganoderma lucidum*. Plant disease. 71 : 251-53.

Almeida Jr. and Aquino, M.de. L.N. 1978. Anthracnose or fruit rot of coconut palm (Cocos nucifera L) in Pemambused State. Persquisa Agropacuaria Pernambucana, 2:37-44.

Alonzo, J.C. and Palomar, M.K. 1980. Effect of seawater and seaweed salt on coconut grey leaf spot diseases. *Philip. Journal of Coconut Studies,* 5 : 27-32.

Amma. P.G.K., Menon, K.S., Patil, P.B. and Sadanandan. 1982. Effect of mixed cropping with Cacao in coconut root (wilt) affected garden. *Indian Cocon.J.* 13 (4) : 14-17.

Anbalagan, R. and Shanmugam, N. 1984. Studies on *Ganoderma lucidum* (Leys) Karst. associated with Thanjavur wilt of coconut. Proc. All India Symp. On Coconut Diseases, Coconut Res. Station, Veppankulam, pp. 24-29.

Anonymous, 1908. Annual Report. Central Plantation Crops Research Institute, Kasaragod.

Anonymous, 1970. Annual Report. Central Plantation Crops Research Institute, Kasaragod.

Anonymous, 1972. Annual Report. Central Plantation Crops Research Institute, Kasaragod.

Anonymous, 1976. Coconut diseases of uncertain etiology. Technical Bulletin No. 1. Central Plantation Crops Research Institute, Kasaragod. pp 22-25.

Anonymous, 1977. Annual Report. Central Plantation Crops Research Institute, Kasaragod. pp 204.

Anonymous, 1978. All India co-ordinator coconut and arecanut improvement project. Progress report 1977-78, CPCRI, Kasaragod. pp. 20-21.

Anonymous, 1978a. All India Co-ordinated Coconut and Arecanut Improvement Project. Progress Report 1975-76, CPCRI, Kasaragod. pp. 14-15.

Anonymous, 1978b. All India Co-ordinated Coconut and Arecanut Improvement Project. Progress Report 1976-77, CPCRI, Kasaragod. pp. 20-21.

Anonymous, 1978c. All India Co-ordinated anc Arecanut Improvement Project. Progress Report 1977-78, CPCRI, Kasaragod. Pp. 22.

Anonymous, 1979. Annual Report. Central Plantation Crops Research Institute, Kasaragod. pp 276

Anonymous, 1980. Annual Report. Central Plantation Crops Research Institute, Kasaragod.

Anonymous, 1981. Report of work done in coconut wilt disease scheme. Coconut Research Station, Veppankulam.

Anonymous, 1981b. Annual Report. Central Plantation Crops Research Institute, Kasaragod.

Anonymous, 1983. All India co-ordinator coconut and arecanut improvement project. Progress report 1981-82, CPCRI, Kasaragod. pp. 57-60

Anonymous, 1985. Annual Report. Agricultural Research. Philippine Coconut Authority, Manila, Philippines.

Anonymous, 1985. All India Co-ordinated Coconut and Arecanut Improvement Project. Progress Report 1984-85, CPCRI, Kasaragod. pp. 72-82.

Anonymous, 1985b. Annual Report. Central Plantation Crops Research Institute, Kasaragod. pp 211

Anonymous, 1986. Coconut root (wilt) disease - Present status of Research and Management - Technical Bulletin 14. CPCRI, Kasaragod, India. pp 10.

Anonymous, 1988. Annual Report. Central Plantation Crops Research Institute, Kasaragod.

Anonymous, 1989. Research Highlights, 1988. Central Plantation Crops Research Institute, Kasaragod, India. pp 12.

Anonymous, 1989a. Annual Report 1988, Central Plantation Crops Research Institute, Kasaragod. pp. 185.

Anonymous, 1990. All India Co-ordinated Research Project on Palms, Project Report 1989. Central Plantation Crops Research Institute, Kasaragod. pp. 61-68.

Anonymous, 1991. Annual Report. Central Plantation Crops Research Institute, Kasaragod.

Anonymous, 1996. Severe bud rot of coconut at Kuttiady area in Kerala. Indian Express. September 20, 1996.

Anonymous, 2000. Annual Report. Central Plantation Crops Research Institute, Kasaragod. pp 115.

Anonymous, 2003. Annual Report. Central Plantation Crops Research Institute, Kasaragod. pp 35.

Anonymous, 2005. Annual Report. Central Plantation Crops Research Institute, Kasaragod.

Ashby, S.F. 1917. Leaf bitten diseases of coconut. Dept. Agric. (Jamaica) Leaflet.

Ashby, S.F. 1920. Notes on two diseases of the coconut palm in Jamaica caused by fungi of the genus Phytophthora. Trop. Agriculturist, 55 : 162-67.

Ashby, S.F. 1921. Report of the Microbiologist for the year 1920. Ann. Rep. Dept. Agric (Jamaica). pp 24-25.

Ashby, S.F. 1925. Bud rot of the coconut palm in the West Indies. *Trop. Agriculturist* 64 : 270-274.

Bachiler, N.C.S.J. and Abad, R.G. 1998. Distribution and disease mapping of stem bleeding disease of coconut (Cocos nucifera L.) in the Philippines. Philippine Journal of Crop Science (Philippines). Vol. 23 : p. 43

Bain, F.M. 1937. Bronze leaf wilt disease of the coconut palm, Govt. Printing Office, Trinidad.

Bakshi, B.K. 1951. Motality of *casuarinas equisetifolia* forst. India for. 77 (4): 269 - 276

Bakshi, B.K., Reddy, M.A.R., Puri, Y.N. and Singh, Sujan. 1972. Forest disease survey. Final Technical Report 1961-1972. Forest Research Institute and Colleges, Dehradun. pp 8.

Baranwal, V.K., Manikandan, P. and Ray, A.K. 1989. Crown choking disorder of coconut : A case of boron deficiency. *Journal of Plantation Crops*. 17 (2) : 114-120

Bavappa, K.V.A., Jayasankar, N.P., Radha, K. and Rethinam, P. 1986. Coconut root (wilt) disease - Present status of research and management. Tech. Bull. 14. CPCRI, Kasaragod.

Bennet, C.P.A., Stepu, G. and Roboth, O. 1985. Aspects of the control of premature nut fall disease of coconut (*Cocos nucifera*) caused by *Phytophthora palmivora* (Butler) Proc. Coconut Plant Protection Seminar, Bogor, 8-10 May, 1985, 29p

Bennett, C.P.A., Roboth, O., Sitepu, G. and Lolong, A. 1986. Pathogenicity of *Phytophthora palmivora* (Butl.) Butl. causing premature nutfall of coconut *(Cocos nucifera L.* Indonesian J. Crop Sci. 2(2) : 59-70.

Bhaskaran, R. 1986. Coconut diseases & their management. In Pest & Disease Management, Oilseeds, Pulses, Millets & Cotton (S. Jayaraj, Ed). TNAU. Coimbatore. pp. 81-89.

Bhaskaran, R. and Ramanathan, T. 1982. Management of Thanjavur wilt of coconut. Proc. PLACROSYM-V. pp. 521-26.

Bhaskaran, R. and Ramanathan, T. 1983. Variability of coconut varities and hybrids to natural infection of gray leaf blight. Madras Agril. J. 70 (10) : 685 – 686.

Bhaskaran, R., Chandrasekar, G. and Jaganathan, T. 1978a. Control of Thanjavur wilt disease of coconut food farming and agriculture, 10 (4) : 115-16.

Bhaskaran, R., Chandrasekar, G. and Jaganathan, T. 1978b. Effect of fertilizers on Thanjavur wilt of coconut. Proc. PLACROSYM-I pp. 351-53.

Bhaskaran, R., Chandrasekar G. and Shanmugam, N. 1985. Problems and priorities in the management Thanjavur wilt of coconut. In Integrated Pest and Disease Management. Proceedings of National Seminar (S. Jayaraj, Ed.). Tamil Nadu Agril. Univ. Coimbatore. Pp. 183-87.

Bhaskaran, R., Ramadoss, N. and Ramachandran, T.K. 1987. Plant extract on the growth of *Ganaderma Lucidum*. Tamil Nadu Agril. Univ. Newsletter, 17 (7) : 2.

Bhaskaran, R., Ramadoss, N. and Suriachandraselvan, M. 1988b. Effect of plant on the growth of *Gandoerma* spp associated with Thanjavur wilt disease of coconut. Proc. of National Seminar on Management of Crop Diseases with Plant Products / Biological Agents. Agril. College & Res. Instt., Madhurai (Abstr.). pp. 24.

Bhaskaran, R., Ramadoss. N. and Ramachandran, T.K. 1988. Biological control of Thanjavur wilt disease of coconut. Indian Coconut J., 19 (6) : 3-8.

Bhaskaran, R., Ramanathan, T. and Ramiah, M. 1982. The Thanjavur wilt. Intensive Agric., 20 (9 & 10) : 19-21.

Bhaskaran, R., Ramanathan, T. and Ramiah, M. 1984a. Thanjavur wilt of coconut : Its occurrence, spread and management. Proc. PLACROSYM - I. Pp. 107-13.

Bhaskaran, R., Ramanathan, T. and Ramiah, M. 1984b. Chemical control of Thanjavur wilt of coconut. Proc. All India Symp. On Coconut Diseases, Coconut Res. Station, Veppankulam, pp., 22-24.

Bhaskaran, R., Rethinam, P. and Nambiar, K.K.N. 1989. Thanjavur wilt of coconut. Journal of Plantation Crops. 17 : 69-79.

Bhaskaran, R., Rethinam, P. and Nambiar, K.K.N. 1994. *Ganoderma* wilt disease of coconut. In Advances in Horticulture Vol. 10. Plantation and Spice Crops Part - 2 (1994) Eds. K.L.Chadha & P. Rethinam, Malhotra Publishing House, New Delhi pp - 858-882.

Bhaskaran, R., Suriachandraselvan, M., and Ramachandran, T.K. 1990. *Ganoderma* wilt disease of coconut - A threat to coconut cultivation in India. *The Planter*, 66: 467-71.

Blacklock, J.S. 1954. A short study of pepper culture with special reference to Sarawak. *Trop. Agriculture 31* (3) : 40-56.

Blair, G. and Darling, D. 1968. Red ring disease on the coconut palm, inoculation studies and histopathology. *Nematologica* 14: 395-403.

Bock, K.R., Ivory, M.H. and Adams, B.R. 1970. Lethal bole rot disease of coconut is East Africa. *Ann. Appl. Biol., 66* : 453-464.

Bose, S.R., 1930. Biology of wood-rotting fungi common in forest areas, *J. Linn.Soc., Lond (Bot).*, 48 : 417 – 38.

Brahmana, J., Lubis, A.U. and Desmier de Chenon. 1992. Evolution of coconut bud rot disease and strategy of control. Coconut *Phytophthora* Workshop Proceedings, October 1992 Manado, *Indonesia, Plant Pathology*. 33-37.

Briton Jones, H.R. 1928. Wilt disease of coconut palm in Trinidad. Part II. *Trop. Agric. Trin. Supplement.* 5 : 1-12.

Briton Jones, H.R. 1940. The disease of the coconut palm. *Bailliere, Tindall and Cox*, London.

Brown, J.F. 1973. Diseases of coconutin the British Solomon Islands. *PL. Dis. Reptr.*, 57 (10) : 856-860.

Bryce, G. 1921. Report on the work of the botanical and mycological division. *Rep. Dept. Agric. Ceylon*, 1920. 13-15 (RAM, 1 : 162)

Bryce, G. 1924b. Coconut Thread Blight (Corticium penicillatum Petch.). *Dept. Agric. New Guinea, Leaflet* 5. (RAM, 4 : 278).

Butler, E.J. 1906. Some diseases of palms. *Agriculture J. India.* 1 : 299-310.

Butler, E.J. 1908. Report on coconut palm disease in Travancore. *Agric. Res. Inst. Pusa. Bull. No. 9* : pp 23.

Butler, E.J. 1909. Fomes lucidus (Leys) Fr., a suspected parasite. *Indian Forester.* 35 : 514-518.

Butler, E.J. 1910. The bud rot of palms in India. *Mem. Dept. Agric. India. Bot. Ser.*, 3 : 221-280.

Butler, E.J. 1926. Some diseases of palms. Agric. J. India : 299-310.

Butler, E.J. and Bisby, G.R. 1913. The fungi of India. Scientific Monograph No. 1. Imperial Council of Agricultural Research, New Delhi. pp 237.

Calica, C.A. and Bigornia, A.E. 1953. Cadang Cadang of Coconut. *Philipp. Agric.*, 37 : 228-240.

Cecil, S.R. 1975. Mineral composition of coconut leaves in relation to root (wilt) disease. *J. Plant. Crops*, 3 : 34-37.

Cecil, S.R. and Amma, P.G.K. 1991. Soils and nutrition. In : Coconut Root (wilt) disease, Monograph Series No. 3. M.K.Nair and Other (Eds), CPCRI, Kasaragod. pp. 45-52.

Chakraborthy, B.K., Nath, B.K., Dhar, P.B. and Gosamy, R.N. 1970. Note on crown rot disease of coconut. *Indian Journal of Agricultural Sciences.* 40 : 502-504.

Chan, C.L.1974. Chemical control of Curvularia leaf spot of coconut seedling. *MARDI Res. Bull.* 2(1) : 19-24.

Chase, A.R. and Broschat, T.K. 1991. Diseases and Disorders of Ornamental Palms. *APS Press pp* 64.

Chettiar, M.R., Varghese, E.J. and Sankaranarayan, M.P. 1959. Chemical studies on the leaf and root (wilt) diseases of Travancore - Cochin V. The effect of lime and ash on disease conditions. Proc. 1st Conf. Cocon. Res. Workers, Trivandrum. India.

Child, R. 1974. Coconut. 2nd edition, Longman Group Ltd., London. pp. 335.

Chona, B.I. and Adansi, A. M. 1970. Coconut in Ghana. Crops Research Institute Bulletin No. 3. Accra, Ghana: Crops Research Institute, Council for Scientific and Industrial Research (CSIR).

Coffe, M.D., Sitepu, D., Kharies, S., Magindan, H., Waroka, J.S. and Motilo, H.F.J. 1990. Integrated control of Phytophthora bud rot and premature nut fall of coconut inn Indonesia. In : Abstract - 3rd Inter. Congr. on Pl. Protn. in Tropics, 20-23 March, 1990, p.105.

Concibidp Manohar, E.C. 1990. Distribution and comparative studies of Phytophthora diseases of coconut in the Philippines. M.Sc. Thesis, University of the Philippines, Los Banos, pp.19-23.

Cooke, F.C. 1950. The tapering disease of coconuts. *Ceylon Coconut Quiart.,* 1 (1) : 17-21.

Cooke, F.C. 1932. Investigations on coconut products. *Dept. of Agriculture, Straits Settlement and Federated Malayan States Bulletin;* 8, 99 pp.

Copeland, E. B. 1931. The Cococnut. Macmillan & Co., Ltd., London.

Dantre, R.K., Verma, S.S., Srivastava, L.K. and Prakash, R. 1997. A survey on bud rot incidence in non-traditional coconut growing area of Eastern Madhya Pradesh. *Indian Coconut Journal* 27: 4-7.

Das, C.M. and Mahanta, I.C. 1985. Evaluation of some fungicides against pestalotia palmarum. CKE incidence of gray blight of coconut pesticides. 19 : 37-39.

Das, T.P.M. and Cheeran, A. 1986. Infectivity of Phytophthora species on cash crops in Kerala. *Agric. Res. J. Kerala,* 24 (1) : 7-13.

Davis, T.A. and Pilli, N.G. 1966. Effect of certain micronutrients on root (wilt) affected and healthy coconut palms in India. *Oleagineux*, 21 (II) : 669-674.

De Leon, D. 1951. Report on studies of a disease (yellow mottled or infections Kadang-Kadang) of the coconut tree in the Philippines- US Mutual Security Agency, Manila.

Deeley, J., Stevans, W.A. and Fox, R.T.V. 1979. 'Use of Dienes' stain to detect plant diseases induced by mycoplasma like organisms. *Phytopathology* 69 : 1169-1171.

Dery, S.K., N'Cho, Y.P., Sangare, A. and Arkhurst E.D. 1997. Cape St. Paul Wilt Disease resistance screening and prospects for rehabilitating the coconut industry in Ghana. Pp. 147- 152, *In*: Eden-Green, S.J. and Ofori, F. (eds). *Proceedings of an international Workshop on Lethal Yellowing-Like diseases of coconut*, Elmina, Ghana, November 1995. Natural Resources Institute, Chatham, UK.

Dwivedi, R.S., Mathew, C., Michael, K.J., Ray, P.K., Sumathy Kuttyamma, B. and Nisan, S. 1978. Carbonic anhydrase carbon assimilation rate and canopy structure in relation to nut on coconut. In Symposium on Crops and Production Physciology, IARI, New Delhi Feb 27, pp 54-56.

Dwivedi, R.S., Mathew, C., Ray, P.K., Amma, B.S.K. and Ninan, S. 1979. Certain closely associated morphological symptoms of root (wilt) disease of coconut (Cocos nucifera). *Plant Disease Reporter*. 63 (6) : 461-463.

Dwyer, R.E.P. 1937. The disease of coconuts (Cocos nucifera) in New Guinea Agricultural Gazette, 3(1) : 28-93.

Fenwick, D.W. 1968. Red ring disease of the coconut palm. In G. C. Smanr, Jr. and V.G.V. Perry (eds.) Tropical Nematology. Center for Tropical Agriculture. University of Agriculture Press, Gainesville.

Frogatt, W.W. 1914. Pests and diseases of the coconut palm. Bull. 2, Dept. Agric., New South Wales.

Furtado, C.X. 1923. Coconut tapering diseases. Trop. Agriculturist, 61: 126. (From RAM, 3 : 269)

Gadd, C.H. 1923. A possible physiological cause of 'nut-fall' of coconuts. *Trop. Agricultrist*, 60 : 112-114.

Gallasch, H.E. 1974. Effect of nutrition on the incidence of *Drechslera incurvata* leaf spot of coconuts. PNG. *Agric. Journal*. 25 : 38-50.

Garofalo, J.F. and Mc Millan, R.T. 1999. *Phytophthora* bud rot of palms in South Florida. Proceeding of the Florida State Horticultural Society, Stuart, Florida, USA. Publ. 2000. No. 112, 110-113.

Garrett, S.D. 1955. A century of root disease investigation. Ann. Appl. Biol. 42 : 211 - 219.

Giddings, N.J. and Wood, J. 1925. Diseases of fruit and nut crops in the US in 1924. Plant Dis. Reptro. Supplement. 39 : 105 (RAM 4 : 608).

Goberdhan, L.C. 1961. Coconut disease in Trinidad and Tobago. 1- Bleeding Stem.J.Agric. Soc. Trin. & Tob. 61 (1):33-38.

Gour, T.B., K.Chandrasekara Rao, D.D.R., Reddy, S.Chander Rao, T. Ramesh Babu, R. Ravindra Babu, M. Sriramulu and P. Narayana Reddy. 2002. Coconut, oil palm - insect pests, diseases, nutritional disorders. Information Technical bulletin No. 17. ANGR Agril. University. Rajendranagar, Hyderabad.

Govindankutty, M.P. 1981. Histopathological studies p. 49-54. In Review of Research of Coconut Root (wilt) disease. CPCRI, Regional Station, Kayamkulam.

Govindu, H.C., Rao, A.N.S. and Keshava Murthy, K.V. 1983. Biology of *Ganoderma lucidum* (Leys.) Karst. And control of Anabe Roga of coconut. In Coconut Research and Development (N.M. Nayar, Ed.).Wiley Eastern Limited. Pp. 325-32.

Gowda, P.V. and Nambiar, K.K.N. 1987. *In vitro* studies on the interaction of antagonistic mycoflora with *Thielaviopsis paradoxa*; the causal agent of coconut stem bleeding disease. M. Phil Dissertation, Mangalore University. Pp. 197.

Griffith, R. 1968. The mechanism of transmission of the red ring nematode. J. Agr. Soc. Trin. Tob. 67: 436-457.

Guest, D. and Drew, R. 2002. Integrated management of Phytophthora disease in tropical horticulture. Proceedings of the international symposium on tropical and subtropical fruits, Cairns, Northern Territory, Australia. 26 November to December 2000. Vol. 2. Acta-Horticulture, 2002. No. 575 (Vol. 2) : 447 - 451.

Gunasekaran, M., Ramdoss, N., Ramiah, M., Bhaskaranm, R. and Ramanathan, T. 1986. Role of neem cake In the control of

Thanjavur wilt of coconut. Indian Coconut J., 17 (1) : 7-12.

Hagley, E.A.C., 1965. The mechanism of transmision of *R. cocophilus* by the palm weevil, *R. palmarum*. Phytopathology 55: 117-118.

Hanold, D., Randles, J.W. 1991a. Coconut Cadang-cadang disease and its viroid agent. Plant Disease. 75 : 330-335.

Harris, D.C., Cardon, J.A., Justun, S.H.F, Passey, W.A.J. 1984. *Phytopthora palmivora* in cultured roots of coconut. Indian Coconut Journal 14:323-330.

Haseloff, J., Mohamed, N.A., Symons, R.H. 1982. Viroid RNAs of Cadang-cadang disease of coconut. Nature 299 : 316-321.

Henry Louis. I, 2002. Coconut - The wonder palm. Hi-Tech Corporation 48/2 Gnana Mutthagam, Ramanputhoor. Nagercoil. pp. 206-218.

Holliday, P. 1980. Fungus diseases of tropical crops. *Cambridge Univ. Press,* Cambridge.

Holmes, F.O. 1965. Report to the Government of India on Investigation on the Etiology of Coconut Root (wilt) disease. Rome, FAO Report No. 1958. Project No. India/TE/PL.

Horne, W. T. 1908. The bud-rot and some other coconut troubles in Cuba. Bull. 15 , Estacion Central Agronomica de Cuba.

Howard, F.W., Norris, R.C. and Thomas, D.L. 1983. Evidence of transmission of lethal yellowing agent by a plant hopper Myndus Crudus (Homoptera : Cixiidae) Trop. Agric. (Trinidad) 60 : 168-171.

Howard, F.W., William, D.S. and Norris, R.C. 1984. Insect transmission of lethal yellowing to young palms. Int. J. Entomol. 26 (331-338.

Imperial, J.S., Bautista, R.M., Randles, J.W. 1985. Transmission of the coconut Cadang-cadang viroid to six species of palm by inoculation with nucleic acid extracts. Plant Pathology. 34 : 391-401.

Indira, P. and Ramadasan, A. 1968. A note on the anatomical derangement in the root (wilt) diseased coconut palm. Curr. Sci. 37 : 290-91.

IRHO - CIRAD Activity Report. 1989-1991. In Oleugineux. 47 : 369-370.

Jackson, G.V.H. and Firman, I.D. 1979. Coconut diseases caused by *Marasmiellus coocphilus* in Solomon Islands. Information circular. South Pacific Commission. No. 83.

Jackson, G.V.H. and Firman, I.D. 1982. Seed-borne marasmioid fungi of coconut - *Pl. Path.*, 31 : 187-188.

Jaganathan, T. and Ramasami, R. 1975. Annual Report 1974-76. Tamil Nadu Agri. Univ., Coimbatore. Pp. 173.

Jaganathan, T. and Ramasami, R. 1977. Pencil point disease of coconut in Tamil Nadu. *Indian Coconut Journal*. 8 : 3-5 & 10.

Jayasankar, N.P. and George, M. 1991. Etiology - B. Bacteria. In : Coconut Root (wilt) disease, Monograph Series No. 3. pp 17-18. CPCRI, Kasaragod.

Joffilly, J.M. 1948. A doenca do and Vermelho do Coquiro e Sua ocorrencia no Brasi Serv. Nac. Pesq. Agron (Rio De Janeiro), 3 : 64.

John, P.C., Channey, P.V. and Verghese, E.J. 1959. 'Chemical studies on the leaf and root (wilt) diseases of coconut in Travancore Cochin VI. Chilean nitrate as a curative for diseased palms and as a fertilizer for coconut seedlings. Proc. First Conf. Cocon. Res. Workers, Trivandrum. India.

Johnson, A. 1965. Host list of fungi etc. and insects recorded in the South East Asia and Pacific Region, *Cocos nucifera* L. Coconut. Technical Report No. 16, FAO Regional Office, Bangkok. pp 18.

Johnston, J.R. 1911. Is Bacillus coli ever a parasite? Phytopathology. 1: 97.

Joseph, K.V. 1983. Phenol metabolism in coconut palms (Cocos nucifera Linn.) in relation to root (wilt) disease. Ph.D thesis, Faculty of Science, University of Kerala.

Joseph, K.V. and Jayasankar, N.P. 1973. Polyphenol content in coconut roots in relation to root (wilt) disease. J. Plant. Crops. 1 (Suppl) : 99-101.

Joseph, K.V. and Jayasankar, N.P. 1979. Phenol metabolism in coconut palms (Cocos nucifera) in relation to root (wilt) disease. Proc. PLACROSYM-II. pp 330-334.

Joseph, K.V. and Shanta, P. 1963. Anatomical changes in the tissues of the tender leaves of coconut palm affected by the root (wilt) disease. J. Indian Bot. Sci., 42 : 61-65.

Joseph, K.V., Potty, V.P. and Jayasankar, N.P. 1976. Increase in polyphenol oxidase and peroxidase with higher intensities of coconut root (wilt) disease. J. Plant. Crops. 4 : 4-6.

Joseph, T. 1978. Some fungi associated with the root system of coconuts in the root (wilt) affected area. Curr. Sci., 47 : 586-587.

Joseph, T. and Lilly, V.G. 1991. Etiology -- A Fungi. In : Coconut Root (wilt) disease. Monograph Series No. 3. p 13-15. CPCRI, Kasaragod.

Julia, J.F. 1982. Mindus taffani (Homoptera : Cixiidae) vector du deperissement foliaine de cocotiers an vanuatu. *Oleagineux* 37 : 409.

Karthikeyan, A. and Bhaskaran, R. 1998. Mineral contents in coconut leaves resistant and susceptible to leaf blight infection caused by pestalotiopsis palmarum. Madras Agricultural Journal. 85 (3 & 4) - 179 - 181.

Karthikeyan, A. and Bhaskaran, R. 1999. Incidence of leaf blight disease in relation to age, vigour of coconut seedlings and yield of palms. Madras Agricultural Journal. 1998 (86) 1 - 3.

Karthikeyan, M., Sarala, L., Karunanithi, K. and Rajarethinam, S. 2002. Control of leaf blight disease of coconut in Tamil Nadu. Indian Coconut Journal. 32 (ii) : 6-7.

Kartika, D., Soebagyo, F.X., Turner, P. D., and Breure, C. J. 1984. Severe stem bleeding and frond fracture of MAWA hybrid coconut in North Sumatra. Paper 51, Intern. Conf. on Cocoa and Coconut, Malaysia. Pp.14.

Kastelein, P. 1987. Observations on red ring disease of coconut palms in Surinam. De Surin. Landb. 33: 40-53.

Khalequzzaman, K.M., Hossain, I. and Hossain, M.M. 1998. Effect of fungicides and potash in controlling grey leaf spot of coconut. Bangladesh Journal of Training and Development. 11 (1 & 2) : 151 - 156.

Khan, E., Seshadri, A.R., Weischer, B. and Mathen, K. 1971. Five new nematode species associated with coconut in Kerala, India. Indian J. Nematol., 1 : 116-127.

Kolandaisamy, S. and Arjunan, G. 1977. Protect coconut trees from diseases. Indian Coconut J., 8 (7) : 3-4.

Koshy, P.K. 2000. Leaf rot disease of coconut. Indian Coconut Journal. 31 (2) : 4-10

Koshy, P.K., Yogesh Kumar, Jayasree, D., Joseph, U., Sosamma, V.K. and Kumar, Y. 2002. Integrated management of leaf rot disease and insect pests on coconut. Indian Phytopathology. 55 (1) : 45-50.

Kumar, P.S. and Hedge, R.K. 1987. Studies on phytophthora diseases of plantation crops. Plant Pathology Newsletter 5:28.

Lal, S.B. 1966. Advances in research on root (wilt) disease problems in Kerala. Proc. Second Sessn. FAO Tech. Wkg. Pty. Prod. Prot. Processg., Bankok, Thailand. pp 273-288

Lee, H.A. 1922. Observations on plant diseases in Philippines. *Phil. Agric. R.,* 14 : 422- 34.

Leefmans, S. 1934. Diseases and pests of cultivated crops in Dutch East Indies in 1931. Meded. Inst. Voor. Plan. Ziekten, 82 :92.

Lewin, H.D., Sindha Mathar, A. and Sethuraman, V. 1983. Effect of fungicides on the control of Thanjavur Wilt disease of coconut. Proc. National Seminar on Management of Diseases of Oilbased Crops. Agril.College & Res. Instt. Madhurai. Pp. 99-100.

Lingaraj, D.S. 1972. Diseases of coconut. Lal-Baugh, 17(3) : 25-31.

Lingaraju, S., Naik, S.T. and Sastry, M.N.L. 1987. Host parasite relationship of Pestalotia palmarum Cooke Plant Pathol. Newsletter, 5 (1-2) : 10.

Mahmud, Z., Zelazny, B. and Sitepu, D. 1990. Phytophthora disease on coconut hybrids in Indonesia. Industrial Crops Research Journal. 2 : 2, 58-61.

Mangindaan, H, Thevenin, J.M. Knarie, S. and Motulo, H.F.J. 1992. The susceptibility of coconut varieties to Phytophthora in Indonesia: The effect of environmental factors. Coconut Phytophthora. Workshop Proceedings. October, 1992, Manodo, Indonesia, pp.39-46.

Mani, K. and Swami, R.N. 1983. Induction of sporulation of *Pestalotiopsis palmarum* by sodium chloride. Trans. Brit. Mycol. Soc. 80 (1) : 151-156

Manjusri Sen, 1973. Cultural Diagnosis of Indian Polyporaceae 3. Genera Daedalea, Favolus, *Ganoderma*, Hexagonia, Irpex, Lenzites, Merulius and Poria. Indian Forest Records, Forest Pathology, 2 (11) : 283-84.

Maramorosch, K. and Kondo, F. 1977. Electron Microscopy of leaf sections from Kerala wilt diseased coconut palm. J. Plant. Crops. 5 : 20-22.

Martyn, E.B. 1934. Bronze leaf wilt of the coconut palm. Proc. Agric. Sco. Trin and Tob., 35 : 507-21.

Martyn, E.B. 1953. Red ring disease of coconut in Trinidad and Tobago. Trop. Agriculture, Trin., 30 : 43-53.

Martyn, E.G. 1945. Coconut diseases in Jamaica (2). Diseases affecting the leaves, crown and stem of coconut. Trop. Agric. 22 (4): 69-76.

Mathai, G. 1980. Seasonal and varietal variations on the incidence of leaf rot of coconut in the root (wilt) affected tracts. Indian Coconut Journal. 11 (6) : 1-3.

Mathen, K. 1969. Investigations on nematodes of coconut. All India Nematology Symposium, Indian Agricultural Research Institute, New Delhi. 1969.

Mathen, K., Kurian, C. and Lal, S.B. 1970. Record of Radopholus similies (Cobb, 1893). Thorne, 1949 and other plant parasitic nematodes from coconut palms, Cocos nucifera L. Sci. Cult., 36 : 159.

Mathen, K., Rajan, P., Radhakirhsnan Nair, C.P., Sasikala, M., Gunasekaran, M., Govindankutty, M.P. and Solomon, J.J. 1990. Transmission of root (wilt) disease to coconut seedlings through Stephanitis typical (Distant) Heteroptera; Tingidae. Tropical Agriculture (Trinidad), 67 (1) : 69-73

Mathew, C. 1977. Changes in carbohydrate content of coconut palm affected by root (wilt) disease. J. Plant. Crops. 5 : 84-88.

Mathew, C., Rajagopal, V., Varkey, T. and Joseph, K.V. 1991. Physiology and Biochemistry. In : Coconut root (wilt) disease. Monograph. Series No. 3. pp 53-59.

Mc Rae, W. 1923. History of the operations against bud-rot of palms in South India. II. Inoculation experiments with P. palmivora Butl. on the Borassus flabellifer Linn. And Cocos nucifera Linn.

Mem. Dept. Agric. India, Bot, Ser., 12 : 21-70.

Menon, K.P.V. 1937. Report on the scheme of investigations on the diseases of coconuts in South India. pp 16.

Menon, K.P.V. and Nair, U.K. 1949. The root (wilt) disease of coconut in Travancore and Cochin. Indian Cocon. J., 3 : 5-10, 40-44.

Menon, K.P.V. and Nair, U.K. 1951. Scheme for the investigation of the root and leaf diseases of coconut palm in South India. Consolidated final report of the work done from 8th March 1937 to 31st March, 1948. Indian Cocon. J. 5 (I) : 3-19.

Menon, K.P.V. and Pandalai, K.M. 1958. The Coconut Palm- A monograph. Indian Central Coconut Committtee, Ernakulam. pp. 384.

Menon, K.S. and Nayar, T.V. 1978. Effect of intercropping with tuber crops in root (wilt) affected coconut garden. Proc. PLACROSYM-I. pp. 416-424.

Michael, K.J. 1964. Studies on the root system of the coconut palm. Indian Cocon. J. 17 : 91-98.

Michael, K.J. 1978. Respiratory rate and nut yield in root (wilt) diseased coconut palms. J. Plant. Crops, 6 : 1-3.

Mishra, D. 1987. New host records of fungi from India. *Indian J. Mycol. & Plant Path.*, 17(2) : 236.

Mitra, M.1 1929. Some diseases of crops in Andaman Islands. Agricultural Research Institute, Pusa, Bulletin 195, pp 14.

Muliyar, M.K., Sri Kailasam. Rawther T.S.S., Pillai. G.B and DAS. P.K. 1982. Coconut package of practices. CPCRI. Pamphlet No. 1 Kasaragod, Kerala. pp., 23

Nagaraj, A.N. and Menon, K.P.V. 1956. Note on the etiology of the root (wilt) disease of coconut palm in Travancore Cochin. Indian Cocon. J. 9 : 161-165.

Nagaraj, A.N., Davis, T.A. and Menon, K.P.V. 1954. Sap transfusion a new device for virus transmission trails in palms. Indian Cocon. J. 7 : 91-98.

Nagarajan, M. 1985. Influence of saline environment on the incidence of stem bleeding in coconut (*Cocos Nucifera L.*). Sci. & Cult, 51 : 349-51.

Naidu, G.V.B., Kumar, S.N.S. and Sannamarappa, M. 1966. Anabe roga, *Ganoderma lucidum* (Leys) Karst on arecanut palm: a review and further observations. Journal of Mysore Horticulture Society. 1 (3) : 14-20.

Naik, R.G. 2001. Chemical control of basal stem rot coconut. Agricultural Science Digest 21(4) : 249.

Nair, M.K.C. and Nathan, H.S. 1971. Spoilage of copra by *Penicillium fraquentas* Westling. Agric.. Res. J. Kerala, 8(2) : 125.

Nair, U.K. and Radha, K. 1962. Manuring cum spraying for the control of coconut palm disease. Proc. First. Confer. Cocon. Res. Workers, India. 1959. pp 250-274.

Nambiar, K.K.N. 1994. Diseases and disorders of coconut. In Advances in Horticulture Vol. X - Plantation and Spice Crops Part - 1 (Eds., Chandha, K.L. and Rethinam, P.) Malhotra Publishing House, New Delhi, pp 857-882.

Nambiar, K.K.N. 2004. Stem bleeding disease. In Advances in the Diseases of Plantation Crops and Spices. Ed. P. Santha Kumari. International Book Distributing Co., Lucknow. pp 173-175.

Nambiar, K.K.N. and Rawther, T.S.S. 1993. Fungal diseases of coconut in the world. In: Advances In Coconut Research and Development. (Eds., Chadha, K.L. and Rethinam, P.), Malhotra Publishing Company Pvt. Ltd., New Delhi, Plant Pathology. 857-882.

Nambiar, K.K.N. and Sastry, K.R. 1988. Stem bleeding disease of coconut. Current status and approaches for control. Philippine J. Coconut Studies, 13 : 30-32.

Nambiar, K.K.N., Anil Kumar, Sastry, K.R. and Joshi, Y. 1989. Seasonal effect on infection by coconut stem bleeding pathogen - *T. paradoxa*. Curr. Sci. 58 : 34-35.

Nambiar, K.K.N., Joshi, Y., Venugopal, M.N. and Mohan, R.C. 1986. Stem bleeding disease of coconut reproduction of symptoms by inoculation with *T. paradoxa*. J. Plant. Crops. 14 : 130-133.

Narasimhachari, C.L. 1990. A report on confirmation of Tatipaka diseased coconut palms in Konaseema of East Godavari district, AP. ARS, Ambajipeta. pp. 21.

Narasimham, M.J. 1921. Report of work done in the Mycological

Section during 1920-21. Admin. Rep. Agric. Dept. Mysore.

Natarajan, S., Bhaskaran, R. and Shanmugham, N. 1986. Preliminary studies to develop techniques for early detection of Thanjavur wilt in coconut. Indian Coconut Journal. 17 (3) : 3 – 6.

Neinhaus, F., Schuiling, M., Gleem, G., Schinzer, U. and Spiltel, A. 1982. Investigations on the etiology of the lethal disease of coconut palm in Tanzania. *Z. pflkrankh pflschutz* 89 : 185-193.

Nishitha Naik, V. 1990. Variability and characteristics of different isolates of *Thielaviopsis paradoxa,* pathogen of coconut stem bleeding disease. M. Phil. Dissertation, Mangalore University. pp. 94.

Nowell, W. 1919. The red ring or "root" disease of coconut palms. *West Indian Bull.* 17: 189-211.

Nowell, W. 1923. Diseases of crop plants in the Lesser Antilles. *The West India Committee,* London.

Ohler. J.G. 1984. Coconut Tree of Life FAO, Rome. pp 446.

Ooka, J.J. and Uchida, J.Y. 1984. Phytophthora Katsurae fruit- rot of Coconut in Hawaii. *Phytopathology* 74:212.

Pandalai, K.M., Sankarasubramoney, H. and Menon, K.P.V. 1954. Sudden wilting of coconut palms at Thottapalli. *Indian Coconut J.* 7 : 107-116.

Pandalai, K.M., Sankarasubramoney, H. and Menon, K.P.V. 1958a. Studies on soil conditions in relation to the root and leaf diseases of the coconut palm in Travancore Cochin. Part IV. Total and exchangeable calcium and magnesium contents of coconut soils. *Indian Cocon. J.,* 11 : 49-56.

Pandalai, K.M., Sankarasubramoney, H. and Menon, K.P.V. 1958b. Studies on soil conditions in relation to the root and leaf diseases of the coconut palm in Travancore Cochin. Part V. Exchangeable cation, cation exchanging capacity and pH of coconut soils. *Indian Cocon. J.,* 11 : 84-101.

Pandu, S.R. and Rajamannar, M. 1983. Symptomatology and indexing of tatipaka disease of coconut. Indian *Phytopathology* 36 (4) : 608- 612.

Papa Rao, A and Govinda Rao, P. 1966. A survey of coconut diseases in Andhra Pradesh. *Andhra Agric. J.,* 13 : 208-17.

Papa Rao, A; Laxminarayan, C and Pandit, V.S. 1975. The nature of coconut leaf blight and its control - *Indian Phytopathology Vol 28,* No. 4, pp. 447-450.

Park, M. 1928. Investigations of root disease of coconuts. *Trop. Agriculturist.* 70 : 402-407.

Park, M. 1932. The effects of drought on coconuts in Puttalam. *Trop. Agriculturist.* 78 : 11-14.

Parthasarathy, M.V. 1978. Have virus like particles been found in leaf of Kerala with diseased coconut palm ? Not yet. *J. Plant. Crops.* 6: 87-89.

Patel, J.S. 1938. The Coconut - A monographs. *Govt. of Madras, Madras,* 313 pp

Patil, S.D. and Kelkar, P.V. 1975. Endosperm rot of palm fruits. *Indian J. Mycol. Plant Path.,* 5:61.

Paul, P.G., Sam Raj, J. and Sushma, P. 1980. Changes in the quality and quantity of coconut oil due to microbial infection of copra. *Agric. J. Kerala,* 18 (1): 68-71.

Peiris, J.W.L. 1962, Bud rot of coconuts. *Ceylon Cocon. Plrs. Review,* 3: 37-40.

Peries, O.S., Liyanage, A. de, S., Mahindapala, R. and Subasinghe, S.M.P. 1975. The incidence of *Ganoderma* root and bole rot of coconut in Sri Lanka. *Ceylon Coconut Quart.,* 26 : 99-103.

Peris, O.S. 1974. Ganoderma basal stem rot coconut : A new record of the disease in Sri Lanka. *Plant Disease Reporter :* 58 (4) 293-295.

Petch, T. 1906. Diseases of the coconut palm. *Tropical Agriculturist,* 27: 489-91.

Petch, T. 1910. Root disease of the coconut palm (Fomes lucidus Leys Fr.). *Circs. & Agric. J. Roy. Bot. Gdns., Ceylon,* 4 : 323-36.

Petch, T. 1916. A preliminary list of Ceylon Polyperi. Ann. Roy. Bot. Gard. *Peradeniya* 6 : 323 – 336.

Pillai, N.G. 1975. Biochemical investigations on the coconut palm. Ph.D. thesis, Faculty of Sciences, University of Kerala.

Pillai, N.G. and Shanta, P. 1965. Free aminoacids in coconut palms affected by root (wilt) disease. *Curr. Sci.,* 34 : 636-637.

Pillai, N.G., Wahid, P.A., Kamala Devi, C.B., Ramanandan, P.L., Cecil, S.R., Kamalakshiamma, P.G., Mathew, A.S. and Nambiar, C.K.B.

1975. Mineral nutrition of root (wilt) affected coconut palm. Proc. Fourth Meeting of FAO Working Party on Coconut Production Protection & Processing, Jamaica. pp.1-20.

Pillai, N.K. 1911. Naleekeram. Vidyabhivardhini Press. Quilon. pp 112

Plavscic-Banjac, B., Hunt, P. and Maramorosch, K. 1972. Mycoplasma like bodies associated with lethal yellowing disease of coconut palm. *Phytopathology*, 62 : 298-299.

Potty, N.N. and Radhakrishnan, T.C. 1978. Stem bleeding disease of coconut-nutrient relationship. Proceedings of PLACROSYM-I : 347-50.

Protacio, D.B. 1964. The *Cerastostomela (Thielaviopsis) paradoxa* leaf spot of coconut new in the Philippines. Philipp. *Journal of Agriculture*. 25 : 67-77.

Quillec, U. and Renard, J.L. 1984. *Phytophthora* rot of coconut. *Oleagineux* 39: 143-147.

Radha, K. 1961. Leaf rot disease of coconut. Proc. First Session FAO Tech. Work. Party, *Cocon. Prod. Prot. and Proc., Trivandrum*. pp 30-36.

Radha, K. 1973. Early warning forecast for coconut bud rot. Annual Report. CPCRI, Kerala.

Radha, K. and Joseph, T. 1983. Bud rot disease of coconut. A reappraisal of associated factors. *In Coconut Research & Development* (Nayar, N.M. Ed). Wiley Eastern Ltd. pp 518.

Radha, K. and Joseph, T. 1974. Investigations on the bud rot disease (*Phytophthora palmivora* Butl.) of coconut. *Final Report, PL* 480 Scheme. Pp. 32.

Radha, K. and Joseph, T. 1976. Studies on bud rot disease of coconut. A reappraisal of associated factors. International Symp. Cocon. Res., Dec. 1976, Kasaragod. (In : Coconut Research and Development Ed.Nayar, N.M.,(1983), Wiley Eastern Ltd., New Delhi, pp. 333-340).

Radha, K. and Lal, S.B. 1972. Diagnostic symptoms of root (wilt) disease of coconut. *Indian J. Agric. Sci.* 42 : 410-413.

Radha, K. and Menon, K.P.V. 1954. Studies on the root (wilt) disease of coconut palm. *Indian Coconut. J.* 7 :99-106.

Radha, K., Sukumaran, C.K. and Prasannakumari, T.O. 1961. Studies on the leaf rot disease of coconut. Fungal infection in relation to environmental conditions. *Indian Coconut Journal.* 15 (1) : 1-11.

Radhakrishnan, T.C. 1987. Symptom expression in stem bleeding diseases of coconut in relation to season. *Indian Phytopath.,* 40 (1) : 100-102.

Radhakrishnan, T.C. 1990. Control of stem bleeding disease of coconut. *Indian Coconut J.,* 20 (9) :13-14.

Radhakrishnan, T.C. and Potti, N.N. 1980. Varietal reaction to coconut stem bleeding disease. *Agric. Res. J. Kerala,* 18 (1) : 118-19

Rahman, M.A., Hossain, M.S. and Choudhury, B.C. 1989. Control of stem bleeding disease of coconut (in Bangladesh). Proceedings of the 6th National Botanical Conference, Chittagong (Bangladesh). Bangladesh Botanical Society, Dhaka. pp 5-6.

Rajagopal, V., Patil, K.D. and Amma, B.S.K. 1986. Abnormal stomatal opening in coconut palms affected with root (wilt) disease. *J. Exp. Bot.,* 37 : 1398-1405.

Rajamannar, M., Krishna Prasadji, J. and Rethinam, P. 1994. Tatipaka disease of coconut. Advances in Horticulture vol. 10 – Plantation and Spice Crops Part-2. (Ed. K.L.Chadha & P. Retinam), Malhotra Publishing House, New Delhil pp.921-937.

Rajan, K.M. 1987. The role of *Ganoderma lucidum* in coconut pathology. International Symposium on *Ganoderma* wilt diseases on palms and other perennial crops. TNAU, Coimbatore (Abstr). pp 6-7.

Rajan, P. and Mathen, K. 1984. Sophonia greeni (Distant) (Nirvanidae: Jassoidea) on leaves of coconut palm, Cocos nucifera L. *J. Plantn Crops,* 12 : 178-179.

Rajan, P. and Mathen, K. 1985. Proutista moesta (Westwood) and other additions to insect fauna on coconut palm. *J. Plantn Crops.* 13 : 135-136.

Ramadasan, A. 1964. Physiology of wilt disease in coconut palm. Proc. 2nd Sessn. FAO Tech. Wkg. Pty. Cocon. Prod. Protc. and Processg., Colombo. pp 257-272.

Ramadoss, N. and Bhaskaran, R. 1987. Effect of fungicides in the management of Thanjavur wilt. International Symposium on *Ganoderma* wilt diseases on palms and other perennial crops.

Tamil Nadu Agrl. Univ., Coimbatore (Abstr.). pp. 12-13.

Ramadoss, N., Gunasekaran, M. and Ramanathan, T. 1985. Effect of neem cake on Thanjavur wilt of coconut. Tamil Nadu Agril University Newsletter, 14 (12) : 1.

Ramadoss, N., Gunasekaran, M. and Ramanathan, T. 1986a. Chemical control of Thanjavur wilt of coconut. In: Seminar on Management of soil-borne diseasese of crop plants (SOMSDOC), Tamil Nadu Agricultural University, Coimbatore (Abstr). pp 21.

Ramanujam, B. 1997. Biocontrol of Thielaviopsis paradoxa (de Seynes) von hohnel, the pathogen of stem bleeding disease of coconut, using antagonistic fungi. Ph.D. thesis submitted to Mangalore University, Mangalore. pp 225.

Ramanujam, B., Nambiar, K.K.N and Rohini Iyer. 2002. Effect of *Gliocladium virens, Trichoderma* spp & organic amendments on the inoculum potential of *T. paradoxa* in soil. Plantation Crops Research and Development in the New Millennium. pp 514-519.

Ramanujam, B., Nambiar, K.K.N. and Anil Kumar. 1993. In : Advances in Coconut Research and Development (Eds. M. K. Nair, H. H. Khan, P. Gopalasundaram and E. V. V. B. Rao). 615-616 Oxford and IBH Publ. Co. Pvt. Ltd., New Delhi, pp 759.

Ramapandu, S., Satyanaryana, Y., Rajamannar, M. and Chiranjeevi, V. 1981. Seasonal variation in the manifestation of *Ganoderma* wilt disease of coconut. *Indian Coconut Journal.* 12 (3) : 5-6.

Ramasami, R., Bhaskaran, R. and Jaganathan, T. 1977. Epidemiology of Thanjavur wilt disease of coconut In Tamil Nadu. *Food Farming and Agriculture,* 9 (6) :147-48.

Randles, J.W. and Hatta, T. 1980. Etiology studies on virus-like diseases of coconut and cardamom in India. Biennial Report of the Waite Agricultural Research Institute, 1978-79. pp. 163.

Randles, J.W., Rodrigueq, M.J.B., Imperial, J.S. 1988. Cadang-cadang disease of coconut palm. *Microbiological Sciences.* 5 : 18-22.

Rao, A.P. and Subramanyam, K. 1975. A new record of Alternaria on coconut leaf (A. alternata). *Indian Phytopath.,* 28, 404-407.

Rao, K.S.N., Sreekantiah, K.R. and Rao, T.N.R. 1971. Post harvest infection of coconut kernel by Botryodiplodia theobromae and a note on the hydrolytic enzymes secreted by fungus. *Indian*

Phytopath., 24 (4) : 815-19.

Rao, P.G., Reddy, G.S. and Subbiah, J. 1956. A new disease of coconuts in Andhra State. *Indian Coconut J.*, 9 : 215-22.

Razin, S. 1983. Identification of mycoplasma colonies. pp 83-88. In methods in Mycoplasmology Vol. I (Eds. Razin, S. and Tully, J.G). Academic Press, New York.

Reinking, O.A. 1961. Yellow mottle decline of coconuts in the territory of Gnam. *Plant Dis. Reptr.*, 45 : 599-604.

Reinking, O.A. 1919. Yellow mottle decline of coconuts in the territory of Guam. *Plant Disease Reporter*, 45 :599-604.

Renard, J.L. and Darwis, S.N. 1993. Report on the Coconut Phytophthora Disease Seminar, Manado, Indonesia, October, 1992. *Oleagineux* 48 (6) : 301-305.

Renard, J.L. Brahmana, D. and Rongnon, F. 1984. Performance of the yellow dwarf x West African tall hybrid coconut with reference to stem bleeding in Indonesia. *Oleagineux*, 39 (6) : 311-319.

Rethinam, P. 1984. Thanjavur wilt disease of coconut in Tamil Nadu. *Indian Coconut J.*, 15 (2) : 3-11.

Rethinam, P., Rajamannar, M. and Narasimhachari, C.L. 1989. Tatipaka disease of coconut in Andhra Pradesh. *Indian Coconut J.*, 20 (1) :1-4.

Rillo, E.P. and Paloma, M.B.F. 1989. Reaction of some coconut cultivars and hybrids to Phytophthora diseases. *Philipp. J. Cocon. Studies* 14 (1) : 2-6.

Rishbeth, J. 1955. Root diseases in plantations with special reference to tropical crops. *Ann. Appl. Boil.* 42 : 220-227.

Rohini Iyer and Rasmi, A.R. 2005. Bud rot disease of coconut in relation to endemic spots in Kerala state, India. Cord, 21 (1) 28-38.

Sahasranaman, K.N., Pillai, N.G., Jayasankar, N.P., Potty, V.P., Varkey, T., Amma, P.G.K. and Radha, K. 1983. Mixed forming in coconut garden - Economics and its effect on root (wilt) diseases. pp 160-165. In Coconut Research and Development (Ed. Nayer, N.M.) Wiley Eastern Ltd., New Delhi. pp 518.

Salas, J.A. 1980. El anillo rojo del cocotero en varias áreas de Costa Rica. *Agron. Cost.* 4: 199-202.

Salgado, M.L.M. 1942. Note on physiological stem bleeding of mature coconut palms. *Trop. Agriculturist.* 98 : 31-35.

Sankarasubramoney, H., Pandalai, K.M. and Menon, K.P.V. 1954. Studies on soil conditions in relation to the root and leaf disease of coconut in Travancore-Cochin. Part. I. Nitrogen, Organic matter content and C/N ratio of coconut soils. Indian Cocon. J. 8 : 5-25.

Sankarasubramoney, H., Pandalai, K.M. and Menon, K.P.V. 1955. Studies on soil condition in relation to the root and leaf diseases of the coconut palm in Travancore Cochin. Part II. Total phosphoric acid, 'Available' phosphoric acid and iron content in coconut soils. *Indian Cocon. J.*, 9 : 20-29.

Sankarasubramoney, H., Pandalai, K.M. and Menon, K.P.V. 1956. Studies on soil condition in relation to the root and leaf diseases of the coconut palm in Travancore Cochin. Part III. Total available and exchangeable potassium content of coconut soils. *Indian Cocon. J.*, 9 : 90-100.

Sasikala, M. and Pillai, N.G. 1978. Transmission of coconut root (wilt) disease - Reaction of 170 test plants. *J. Plant. Crops,* 6 (1) : 17-20.

Sasikala, M., Mathen, K., Govindankutty, M.P., Solomon, J.J. and Geetha, L. 1988. Transmission of mycoplasma like organism from Cocos nucifera with root (wilt) disease to Catharanthus roseus by Cassytha filiformis. *Neth. J. Pl. Path.,* 94 : 191-194.

Satyanarayana, Y., Ramapandu, S., Rajamannar, M. and Chiranjeevi, V. 1985. Control of *Ganoderma* wilt disease of coconut. *Indian Coconut J.,* 16(5) : 3-5.

Schuiling, M., Kazia, D.A. and Mpunami, A. 1992. Lethal disease of coconut palm in Tanzania IT. History, distribution and epidemiology. *Oleagineux* 16-521.

Seemuller, E. 1976. Investigations to demonstrate mycoplasma like organisms in diseased plants by fluorescence microscopy. *Acta Hort.* 67 : 109-112.

Shanta, P. and Menon, K.P.V. 1960. Cowpea (Vigna sinensis Endl.), an indicator plant for the coconut wilt virus. *Virology,* 12 : 309-310.

Shanta, P. and Menon, K.P.V. 1961. Studies on some properties of the coconut wilt virus. *Indian Coconut J.,* 15 : 36-46.

Shanta, P., Hariharasubramanjan, V., and Pillai, N.G. 1975. Possible association of tobacco mosaic virus with the root (wilt) disease of coconut. *J. Plantn. Crops.* 3 : 77-80.

Shanta, P., Joseph, T. and Lal, S.B. 1964. Transmission of root (wilt) disease in coconut. *Indian Cocon. J.* 18 : 25-28.

Sharma, R.B., Roy, A.N. and Sharma, M.P. 1985. A new disease of coconut kernel. *Curr. Sci.,* 54 : 199-200.

Sharples, A. 1925. Observations on the bud rot of palms. Rep. Imp. Bot. Conf., London, 1924. *Trop. Agriculturist,* 64 : 257-277.

Sharples, A., 1923. Annual Report of the Mycologist for the year 1922. '26, '27 and '29. *Malay. Agric. J.,* 11 : 267-72; Ibid., 14., 65-73; ibid., 15 : 152-59; ibid., 17:294-303.

Shaw, D.E. and Booth, C. 1967. Petiole disease of coconut in Papua. Papua New Guinea *Agric. J.,* 19: 89-93.

Shepherd, E.F.S. 1926. Botanical Division. *Ann. Rep., Mauritius Dept., Agric.,* 15-18.

Simmonds, H.W. 1924. The diseases of coconuts (*Cocos nucifera*) in New Guinea. New Guniea Agric. Gaz., 3 : 28-93.

Sindha Mathar, A. and Balasubramaniam, M. 1987. Control of Thanjavur wilt disease of coconut . International Symposium on *Ganoderma* Wilt Diseases on Palms and other Perennial Crops. Tamil Nadu Agril. Univ. Coimbatore (Abstr.). pp. 11-12.

Sindha Mathar, A., Lewin, H.D. and Sethuraman, V. 1983. Effect of systemic and non-systemic fungicides on the development of Thanjavur wilt disease of coconut. Proc. National Seminar on Management of Diseases of Oilseeds Crops. Agri. College & Res. Instt. Madhurai. Pp. 98-99.

Small, W. 1927. Matters of Phytopathological interest during 1926. Year Book, Dept. Agric. Ceylon : 7-10.

Solomon , J.J. and Govindankutty, M.P. 1991. Etiology E. Mycoplasma like organisms. In Coconut Root (wilt) disease. Monograph Seried No. 3. CPCRI, Kasaragod. India. pp 92.

Solomon, J.J. 1991. Recent advances in research on root (wilt) disease of Coconut. J. Plantn. Crops. 19 (2) : 153-162.

Solomon, J.J. 1997. Phytoplasma diseases of coconut - Root (wilt) and Tatipaka diseases. In Summer Institute Publication on 'Recent

Advances in the Integrated Management of Pests and Diseases of coconut' conducted at CPCRI, Kayangulam during 10-30 October 1996. pp. 25-43.

Solomon, J.J. and Geetha, L. 1993. Studies on the Ultrastructural changes in root (wilt) diseased palms. J. Plantn. Crops. 21 (Suppl.): 113-118.

Solomon, J.J., Govindankutty, M.P. and Neinhaus, P. 1983. Association of mycoplasma like organisms with coconut root (wilt) disease in India. Z. pflkrankh pflschutz. 90 : 295-297.

Solomon, J.J., Nair, C.P.R., Srinivasan, N., Gunasekaran, M. and Sasikala, M. 1999. Coconut root wilt - the malady and remedy. Journal of Plantation Crops. 27 (2) : 71-92.

Sosamma, V.K. 2001. Nematodes associated with spindle leaves of leaf sot affected coconut palms. Indian Journal of Nematology. 32 (2) : 167-168.

Sosamma, V.K. and Koshy, P.K. 1978. A note on the association of Cylindrocarpon spp. with Radopholus similes in coconut. Indian Phytopath., 31 : 381-382.

Sosamma, V.K. and Koshy, P.K. 1991. Etiology - Nematodes in coconut root (wilt) disease. Monographs Series No. 3 p. 19-26. CPCRI, Kasaragod.

Srinivasan, N. 1998. 'Coconut leaf rot : Research Accomplishments since 1990'. Group Meeting on root (wilt) and leaf rot disease of coconut. CPCRI Regional Station, Kayangulam, Kerala. April 21-22, 1998.

Srinivasan, N. 1999. Know about phytoplasma as pathogen in relation to coconut diseases and with reference to root wilt. Indian Coconut Journal. 30 (7) : 3-6.

Srinivasan, N. 2001b. Diseases and disorders of coconut and their management. In: Plant Pathology (Ed: Trivedi, P.C.), Pointer Publishers, Jaipur, India, Plant Path. 194-25.

Srinivasan, N. 2002a. Coconut leaf rot complex and perspectives for the disease control status report. *Indian Coconut Journal* 32 (9) : 2-9.

Srinivasan, N. 2002d. Coconut leaf rot complex and perspectives for the disease control - Status report (Reproduced). *The planter* 78

(913) : 203-216.

Srinivasan, N. and Gunasekaran, M. 1992. An appraisal of symptom expression in coconut due to leaf rot disease. *Indian Coconut Journal.* 23 (7) : 2-6.

Srinivasan, N. and Gunasekaran, M. 1996a. Field control of leaf rot disease of coconut with fungicides. *CORD.* 12 (2) : 34-42.

Srinivasan, N. and Gunasekaran, M. 1998a. Etiology and recurrence of coconut leaf rot with special reference to seedlings. In : Abstracts of papers, PLACROSYM-XIII, Coimbatore, India, December 16-18, 1998 (Ed: Muraleedharan, N), UPASI, Valparai, India, pp 54-55.

Srinivasan, N. and Gunasekaran, M. 2000a. Etiology and recurrence of coconut leaf rot with special reference to seedlings. In : Recent Advances in Plantation Crops Research'. PLACROSYM-XIII, Coimbatore, India, December 16-18, 1998 (Ed: Muraleedharan, N and R. Raj Kumar), Allied Publishers Ltd., New Delhi. pp 400-403.

Srinivasan, N. and Gunasekaran, M. 1995. Additional fungi associated with leaf rot disease of coconut in India (A research note). *The Philip.J. of Coconut Studies* XX(2) : (in press)

Srinivasan, N. and Rohini Iyer. 2006. Leaf rot and bud rot of coconut. In : Advances in the Diseases of Plantation Crops and Spices (Ed. P. Santha Kumari). *International Book Distribution Company.* pp 122-152

Srinivasan, N., Gunasekaran, M. and Alka Gupta. 2002. Substantive evidence of complex fungal etiology of coconut leaf rot disease. *Indian Coconut Journal 32* (12) : 12-15.

Srinivasan, N., Gunasekaran, M., Joseph, T. and Rawther, T.S.S. 1998. Leaf rot disease. In : Coconut root (wilt) disease (Eds. Nampoothiri, K.U.K. and Koshy, P.K.) CPCRI, Kasaragod, India. pp 83-95.

Srinivasan, N., Gunasekaran, M., Rohini Iyer, Chandramohanan, R. and Kochu Babu, M. 1995. Mycoflora of rot affected coconut palms. *Indian Phytopathology* 48 (1) : 93-95.

Srinivasan, N., Solomon, J.J., Sasikala, M., Gunasekaran, M., Geetha, L. and Rajeev, G. 2001. Managing coconut root wilt and leaf rot diseases. Indian Horticulture 46 (3) : 28-29.

Srinivasulu, B., Aruna, K., Sabitha Doraisamy and Rao, D.V.R. 2001a. Occurrence and biocontrol of *Ganoderma* wilt disease of coconut in coastal agro ecosystem of AP. *J. Indian Soc. Coastal Agric. Res.* 19 (1 & 2) : 191-195.

Srinivasulu, B., Aruna, K. and Rao, D.V.R. 2001b. Biocontrol of *Ganoderma* wilt of coconut palm. *South Indian Hort.* 49 (Special) : 240-243.

Srinivasulu, B., Aruna, K., Krishna Prasadji, J., Rajamannar, M., Sabitha Doraisamy, Rao, D.V.R. and Hameed Khan, H. 2002a. Prevalence of basal stem rot disease of coconut in coastal agro ecosystem of AP. *Indian Coconut Journal XXXIII* (Nov, 2002) : 23-26.

Srinivasulu, B., Sabitha Doraisamy, Aruna, K., Rao, D.V.R. and Rabindran, R. 2002b. Efficacy of biocontrol agent, chemicals and botanical on *Ganoderma* spp, coconut basal stem rot pathogen. *Journal of Plantation Crops.* 30 (3) : 57-59.

Srinivasulu, B., Aruna, K., Lakshmi, B.K.M., Sabitha Doraisamy, Rao, D.V.R. and Hameed Khan, H. 2002c. *T. hamatum* - A potential biocontrol agent for basal stem rot (*Ganoderma* wilt) disease of coconut. *Proc. of PLACROSYM-XV* (2002) : 541-544.

Srinivasulu, B., Aruna, K., Rao, D.V.R. and Hameed Khan, H. 2003a. Epidemiology of basal stem rot (*Ganoderma* wilt) disease of coconut in AP. *Indian Journal of Plant Protection.* 31 (1) : 48-50.

Srinivasulu, B., Rajamannar, M., Aruna, K., Rao, D.V.R. and Khan, H.H. 2003b. Sero detection of coconut basal stem rot pathogens–*G. applanatum* and *G. lucidum*. *Journal of Plantation Crops.* 31 (2) : 62-64.

Srinivasulu, B., Aruna, K., Vijay Krishna Kumar, K. and Rao, D.V.R. 2003c. Investigations of post harvest aflatoxin contamination of copra. *Indian Coconut Journal.* 34 (8) : 8-9.

Srinivasulu, B., Vijay Krishna Kumar, K., Aruna, K., Lakshmi, B.K.M. and Rao, D.V.R. 2004a. An ecofriendly technology to manage basal stem rot (*Ganoderma* wilt) disease of coconut. *The Andhra Agric. Journal* 50 (Spe) : 264-267.

Srinivasulu, B., Vijay Krishna Kumar, K., Aruna, K. and Rao, D.V.R. 2004b. Biocontrol of major pathogens of coconut. *Journal of Plantation Crops.* 32 (Suppl) : 309-313.

Srinivasulu, B., Vijay Krishna Kumar, K., Aruna, K., Sabitha Doraisamy and Rao, D.V.R. 2004c. Biocontrol potentiality of *Trichoderma viride* against basal stem rot disease of coconut. *Journal of Plantation Crops.* 32 (1) : 28-31.

Srinivasulu, B., Vijay Krishna Kumar, K., Kochu Babu, M., Satyavani, V. and Aruna, K. 2005a. An overview of basal stem rot of oilpalm and BSR-IDM in coconut. National Seminar on Research and Development of Oil Palm in India. NRC for oil palm, Pedavegi, West Godavari Dist., AP.

Srinivasulu, B., Krishna Prasadji, J., Rajamannar, M., Vijay Krishna Kumar, K., Aruna, K. and Rao, D.V.R. 2005b. Basal stem rot (*Ganoderma* wilt) of coconut. *Technical Bulletin,* AICRP on Palms, Agricultural Research Station, ANGRAU, Ambajipeta.

Srinivasulu, B., Vijay Krishna Kumar, K., Aruna, K., Lakshmi, M.V. and Rao, D.V.R. 2006a. A novel approach in the management of stem bleeding disease of coconut. 93rd Indian Science Congress at ANGRAU, Hyderabad. 3-7 Jan, 2006.

Srinivasulu, B., Vijay Krishna Kumar, K., Rajamannar, M., Krishna Prasadji, J., Rao, D.V.R. Lakshmi, M.V. and Aruna, K. 2006b. Stem bleeding disease of coconut. *Technical Bulletin,* AICRP on Palms, Agricultural Research Station, ANGRAU, Ambajipeta.

Srinivasulu, B. 2006c. Recent advances in coconut disease management. Seminar on Production, Processing and Marketing of Coconut' organized by National Coconut Growers Federation & Coconut Development Board. 24th January, 2006.

Srinivasulu, B. 2006d. Ecofriendly management of plant diseases. National Seminar on Environment and Development at Centre for Biotechnology, Acharya Nagarjuna University, Guntur. 30th June, 2006.

Srinivasulu, B., Vijay Krishna Kumar, K., Aruna, K. and Rao, D.V.R. 2006e. Biosuppression of *Aspergillus Flavus,* a mycotoxicant of copra. *Journal of Plantation Crops.* Vol. 34 (1)

Srinivasulu, B., Lakshmi, M.V. and Aruna, K. 2006f. *In vitro* effects of physical, chemical and biological factors on *Phytophthora palmivora. AAPP News Letter Vol* I (2) : 4.

Srivastava, D.N., Shekhawat, G.S. and Rao, Y.P. 1969. Association of bacteria with root (wilt) of coconut in Kerala. *Indian J. Agric. Sci.,*

39 : 395-382.

Steer, J. and Coates-Beckford, P.L. 1990. Role of phytophthora Kasturae, P. palmivora, *Thielaviopsis paradoxa* and Enterobacter in bud rot disease of coconut in Jamaica. *Oleagineux 45* (12) : 539-545.

Steyaert. R.L. 1948. The root rot problem in the tropical culture of woody plants and its bearing on methods of opening up plantations. *Bull. Agric. Cango belge.* 40 (2) : 1651-1678.

Stockdale, F.A. 1921. Pests and diseases. *Trop. Agriculturist,* 56-94.

Stockdale, F.A. 1907. Coconut palm disease in Trinidad. *Bull. Dept. Agric. Jamaica* 5 : 111-139.

Subbiah, J. and Rao, P.G. 1963. Studies in coconut decline in Andhra. *Andhra Agric. J.,* 10 : 58-67.

Subramanyam, V. 1965. Control of infection and production of quality products from coconuts without machinery or application of heat. *Philipp. J. Nutr., 218-34.*

Sulladmath, V.V. and Ponnappa, K.M. 1978. Petiole disease of coconut palm caused by *Anthostomella fusispora* Booth & Shaw. *Journal of Plantation Crops,* 6(2) : 65-66.

Sulladmath, V.V. and Shantappa, P.B. 1980. Aetiology of stem bleeding disease of coconut in semi-dry tract of Karnataka. *Cur. Res.,* 9 (5): 73-75.

Sulladmath, V.V. and Shanthappa, P.B. 1979. Nut fall in coconut associated with *Botryosphaeria rodina* (Berk. And Curt.) V. *Arx. Curr. Res.,* 8:13.

Summanwar, A.S., Raychaudhuri, S.P., Jagadish Chandra, K., Namaprakahsh and Lal, S.B. 1969. Virus associated with coconut root (wilt) disease. *Cur. Sci.* 38: 208-210.

Sundararaman, S. 1922. The coconut bleeding disease. *Bulletin of Agricultural Research Institute,* Pusa. Pp. 127.

Sundararaman, S. 1925. Preliminary note on coconut leaf rot of Cochin. Year Book, *Madras Agricultural Department,* 1924. pp 6-8.

Sundararaman, S. and Ramakrishnan, T.S. 1924. The 'Mahali' disease of coconuts in Malabar. Memoirs of Dept. of Agriculture, India. Botanical Series, 13 : 87-97.

Suriachandraselvan, M. and Bhaskaran, R. 1999. Management of Ganoderma / Thanjavur wilt of coconut by non-chemical methods. *Indian coconut J.* 30 (8) : 4-5.

Susamma, P. and Menon, M.R. 1983. Storage studies of copra. Proceedings of the National Seminar on Management of Diseases of Oilseeds Crops, Tamil Nadu Agricultural University, Madurai, India. pp 107.

Susamma, P. 1980. Production of cellulolytic enzymes by the fungi associated with the spoilage of copra. *Madras Agricultural J.*, 67:603-605.

Tey, C.C and Chan, E. 1980. Diseases of coconut palms in Peninsular Malaysia. Proc. of the International Conf. on Cocoa, Coconuts. Kuala Lumpur, Malaysia. 21-24 June 1978. pp 692-707. MARDI.

Thompson, A. 1924. Annual Report of the Mycologist for 1923. Malay. Agric. J., 12 : 246-251. Notes on various diseases (RAM, 4 : 84).

Thompson, A. 1933. Division of Mycology, Annual Report for 1931. Department of Agriculture , Strait Settlement and Federated Malay States Bulletin. 12:48-52.

Tsai, J.H. 1982. Lethal yellowing of palms. In : Plant diseases of viral, viroid, mycoplasma and uncertain etiology. (Ed. Karl Maramorosch) pp 175.

Tucker, C.M. 1926. Phytophthora bud rot of coconut palms in Puerto Rico. *Journal of Agricultural Research,* 32 : 471-98.

Turner, P.P., Jones, P. and Kenten, R.H. 1978. Coconut stem necrosis, a disease of hybrid and Malayan dwarf coconuts in North Sumatra and Peninsular Malaysia. Perak. *Planters. J.*, 33 : 33-46.

Tuset, J.J., Hinare Jos, C., Buj, A, Molins, A. and Cebolla, V. 1992. Myrothesium roridum and Mverrucaria, as potential antagonists of Phytophthora Spp. on coconut. Coconut Phytophthora proceedings, 26-30 October 1992. Manodo, Indonesia, pp. 171-176.

Tymon, A.M., Jones, P. and Eden-Green S.J. 1995. I: Detection and discrimination of mycoplasma like organisms (MLO) associated with coconut lethal yellowing and other diseases. *N.R.T. Project A 0882. Final Technical Report R 2970(s).* 21pp.

Tymon, A.M., Jones, P. and Harrison, N.A. 1998. Phylogenetic relationships of coconut phytoplasmas and the development of specific oligonucleotide PCR primers. Annals of *Applied Biology* 132: 437-452.

Uchida, J.Y., Ooka, J.J., Nagata, M. and Kadooka, C.Y. 1992. A new Phytophthora fruit and heart rot of coconut in Hawaii. *Plant Disease* 76 : 925-927.

Usman, N.M. 1988. Studies, on stem bleeding disease of coconut. M. Phil dissertation, Chlorine Mangalore University. pp. 74.

Van Overem, C. 1926. Over het voorkomeri van *Ganoderma lucidum* (Leysser) Karst. in rubbertuinen. Arch. V. Rubbercult. Nederl. Ind., 9 : 518-521.

Varkey, T. and Davis, T.A. 1960. Studies on coconut pollen with reference to the leaf and root (wilt) diseases. *Indian Cocon. J.* 14 : 1-7.

Varkey, T., Michael, K.J. and Ramadasan, A. 1969. Note on the investigation on the nitrogen and phosphorus metabolism of coconut palm affected by root (wilt) disease. *Indian J. Agric. Sci.,* 39 : 25-26.

Veena, S.S., Peethambaran, C.K, and Sarma, Y.R. 1997. Phytophthora Katsurae as the causal agent of bud rot of coconut in some part of Kerala. Abstracts of Symposium and Poster Sessions, International Conference on Integrated Plant Disease Management For Sustainable Agriculture, IPS, ICAR & IARI, New Delhi, Nov. 1997, pp.272.

Venkatarayan, S.V. 1935, Control of 'Anaberoga' of areca and coconut palm. *Mysore Agric. Cel.* 1935

Venkatarayan, S.V. 1936. The biology of *Ganoderma lucidum* on areca and coconut palms. *Phytopathology* 26 : 153-75.

Verghese, E.J., Sankaranarayanan, M.P. and Menon, K.P.V. 1962a. Chemical studies on the leaf and root disease of coconut in Travancore Cochin II. Nutrient content of healthy and diseased palm. Proceedings of the First Conference of Coconut Research Workers, India. 1959. pp 310-319. Indian Central Coconut Committee, Ernakulam, Kerala.

Verghese, E.J., Sankaranarayanan, M.P. and Menon, K.P.V. 1962b. Chemical studies on the leaf and root disease of coconut in Travancore Cochin III. Cadmium content of coconut roots and leaves. Proceedings of the First Conference of Coconut Research Workers, India. 1959. pp 366-370. Indian Central Coconut Committee, Ernakulam, Kerala.

Verghese, M.K. 1934. Diseases of the coconut palm. Government Press, Trivandrum. pp 105.

Vijayan, K.M. and Natarajan, S. 1972. Some observations on the coconut wilt disease of Tamil Nadu. *Coconut Bull*, 2 (12) : 2-4.

Vijayan, K.M. and Natarajan, S. 1975. Influence of fertilizer and manuring on the incidence and progress of coconut wilt disease of Tamil Nadu. *Coconut Bull.*, 5 : 1-5.

Vijayan, K.M., Natarajan, S. and Krishnamoorthy, C.S., 1973. Coconut wilt disease of Tamil Nadu. *Madras Agric. J.*, 60 : 504-506.

Vijayaraghavan, H., Ramadoss, N., Ramanathan, T. and Rethinam, P. 1987. Effect of toddy tapping on Thanjavur wilt disease of coconut. *Indian Coconut Journal.* 17 (12) : 3-5.

Vijayaraghavan, H., Ramadoss, N., Ravendran, T.S. and Ramanathan, T. 1986. Management of Thanjavur wilt of coconut by tapping . Seminar on Management of Soil-borne Diseases of Crop Plants. TNAU, Coimbatore (Abstr). pp. 36.

Von Uexkull, H.R. 1985. Chlorine in the nutrition of palm trees. *Oleagineux*, 40 : 67-74.

Ward, F.S. 1937. Insects, fungi and bacteria associated with copra in Malaya. 2. Deterioration of copra caused by bacteria and moulds. Pp. 108. Dept. of Agr. St. Settlements and FMS.

Warwick, D.R.N. and Abakerli, R.B. 2001. Chemical control of lixas and leaf blight disease of coconut palms. 45(4) : 168-170.

Weischer, B. 1967. Report to the Government of India on plant parasitic nematodes. Report No. TA 2332. UVN, Rome.

Wijesekara, H.T.R., Rajapakse, C.N.K., Fernando, L.C.P, Jaya Singhe, C.K., Manoher, S.P. and Perera, K.F.G. 1999. Cocos. 13 : 21 – 29.

Wilson, K.I., Rajan, K.M., Nair, M.C. and Balakrishnan, S. 1987. *Ganoderma* disease of coconut in Kerala. International Symposium on *Ganoderma* Wilt Diseases on Palms and other Perennial Crops. TNAU, Coimbatore (Abstr.). pp. 4-5.

Zelazny, B., Randles, J.W., Boccardo, G. Imperial, J.S. 1982. The viroid nature of the Cadang-cadang disease of coconut palm. *Scientia Filipnas.* 2 : 45-63.

Plate-1: *Ganoderma* infected coconut roots

Plate-2: Basal stem rot disease : Stem bleeding symptom

Plate-3: Basal stem rot disease : Dead palm

Plate-4: *Ganoderma* spp sporophores

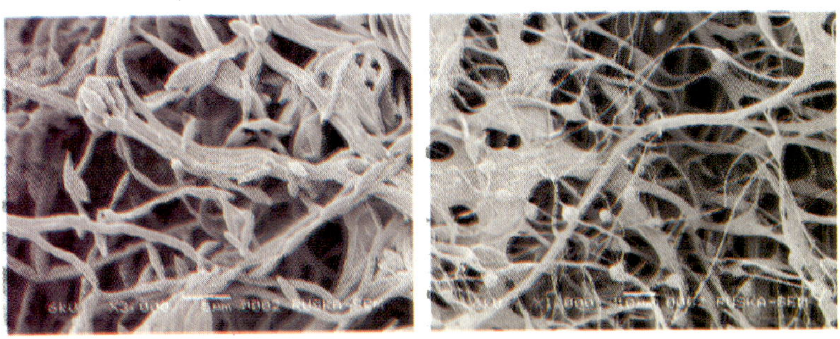

G. applanatum   *G. lucidum*

Plate-5: Scanning Electron Microscopy of *Ganoderma* spp

Plate-6: Bud rot symptoms on Crown

Plate-7: Bud rot symptoms on Nut

Plate-8: Bud rot affected coconut palm

Plate-9: Stem bleeding disease symptoms on coconut palm

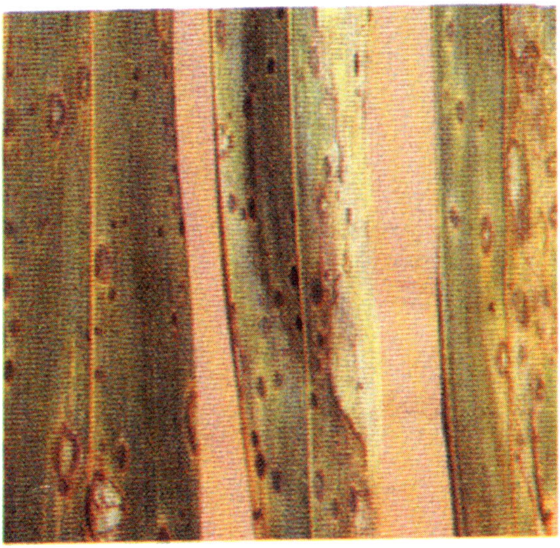

Plate-10: Leaf blight : Greyish-white spots surrounded by brown bands

Plate-11: Leaf blight : affected leaves with irregular necrotic patches

Plate-12: Leaf rot : Blackening and shriveling of the distal ends of the leaflets

Plate-13: Leaf rot : Severe leaf rotting symptoms on newly formed leaves

Plate-14: Root (wilt) disease of coconut

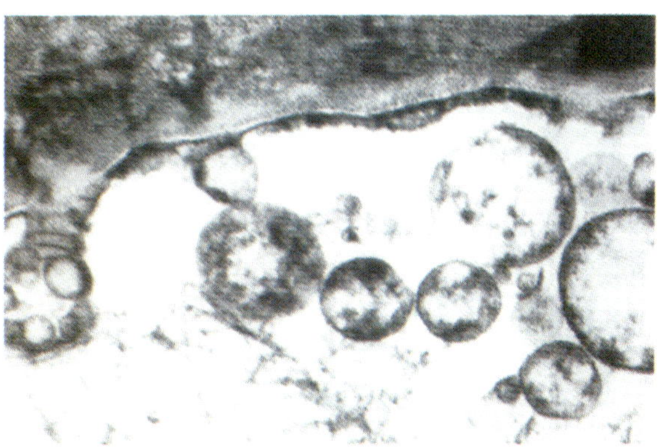

Plate-15: Phytoplasma - Root (wilt) disease of coconut

Plate-16: Tatipaka disease of coconut

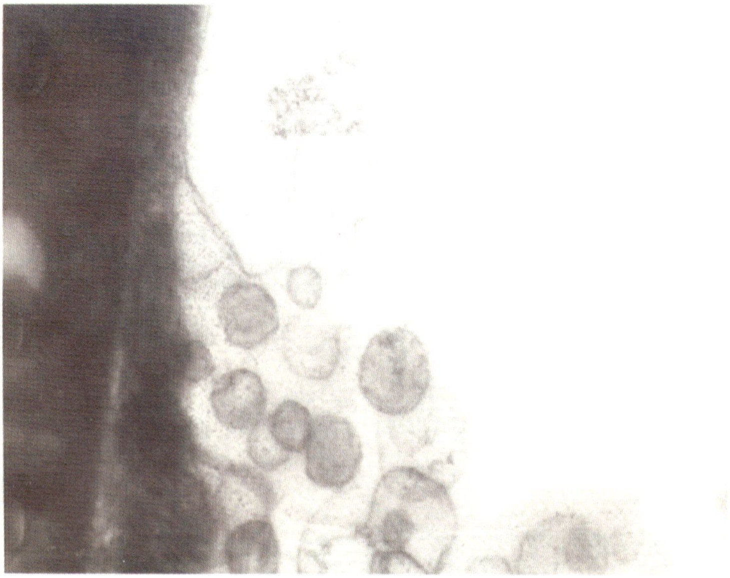

Plates-17: Phytoplasma - Tatipaka disease of coconut

Index